SpringerBriefs in Applied Sciences and Technology

T0172177

SpringerBriefs present concise summaries of cutting-edge research and practical applications across a wide spectrum of fields. Featuring compact volumes of 50–125 pages, the series covers a range of content from professional to academic.

Typical publications can be:

- A timely report of state-of-the art methods
- An introduction to or a manual for the application of mathematical or computer techniques
- A bridge between new research results, as published in journal articles
- A snapshot of a hot or emerging topic
- An in-depth case study
- A presentation of core concepts that students must understand in order to make independent contributions

SpringerBriefs are characterized by fast, global electronic dissemination, standard publishing contracts, standardized manuscript preparation and formatting guidelines, and expedited production schedules.

On the one hand, **SpringerBriefs in Applied Sciences and Technology** are devoted to the publication of fundamentals and applications within the different classical engineering disciplines as well as in interdisciplinary fields that recently emerged between these areas. On the other hand, as the boundary separating fundamental research and applied technology is more and more dissolving, this series is particularly open to trans-disciplinary topics between fundamental science and engineering.

Indexed by EI-Compendex, SCOPUS and Springerlink.

More information about this series at http://www.springer.com/series/8884

Diego T. Santos · Ádina L. Santana ·
M. Angela A. Meireles · Ademir José Petenate ·
Eric Keven Silva · Juliana Q. Albarelli ·
Júlio C. F. Johner · M. Thereza M. S. Gomes ·
Ricardo Abel Del Castillo Torres ·
Tahmasb Hatami

Supercritical Antisolvent Precipitation Process

Fundamentals, Applications and Perspectives

 Springer

Diego T. Santos
LASEFI/DEA, School of Food Engineering
University of Campinas—UNICAMP
Campinas, São Paulo, Brazil

M. Angela A. Meireles
LASEFI/DEA, School of Food Engineering
University of Campinas—UNICAMP
Campinas, São Paulo, Brazil

Eric Keven Silva
LASEFI/DEA, School of Food Engineering
University of Campinas—UNICAMP
Campinas, São Paulo, Brazil

Júlio C. F. Johner
LASEFI/DEA, School of Food Engineering
University of Campinas—UNICAMP
Campinas, São Paulo, Brazil

Ricardo Abel Del Castillo Torres
LASEFI/DEA, School of Food Engineering
University of Campinas—UNICAMP
Campinas, São Paulo, Brazil

Ádina L. Santana
LASEFI/DEA, School of Food Engineering
University of Campinas—UNICAMP
Campinas, São Paulo, Brazil

Ademir José Petenate
Process Improvement
EDTI
Campinas, Brazil

Juliana Q. Albarelli
LASEFI/DEA, School of Food Engineering
University of Campinas—UNICAMP
Campinas, São Paulo, Brazil

M. Thereza M. S. Gomes
LASEFI/DEA, School of Food Engineering
University of Campinas—UNICAMP
Campinas, São Paulo, Brazil

Tahmasb Hatami
LASEFI/DEA, School of Food Engineering
University of Campinas—UNICAMP
Campinas, São Paulo, Brazil

ISSN 2191-530X   ISSN 2191-5318   (electronic)
SpringerBriefs in Applied Sciences and Technology
ISBN 978-3-030-26997-5   ISBN 978-3-030-26998-2   (eBook)
https://doi.org/10.1007/978-3-030-26998-2

This Springer imprint is published by the registered company Springer Nature Switzerland AG
The registered company address is: Gewerbestrasse 11, 6330 Cham, Switzerland

Introduction

The use of supercritical fluids has emerged in different fields. Supercritical fluid technologies to precipitate target compounds offer several advantages over conventional ones, such as low energy requirements, low thermal and chemical degradation of products, and the production of solvent-free particles with narrow size distributions. Microparticles and nanoparticles can be formed directly from substance solutions in a single-step supercritical fluid process based on its use as an antisolvent. This could be an excellent alternative since milling, grinding, and crushing process can lead to contamination of the product, batch variation, downstream processing difficulties, degradation of heat-sensitive materials during grinding, chemical degradation due to exposure to the atmosphere, long processing times, and high energy consumption. Thus, this book provides deep insights into the fundamentals, applications, and perspectives of the supercritical antisolvent (SAS) precipitation, and correlated processes.

Chapter 1, entitled "A Detailed Design and Construction of a Supercritical Antisolvent Precipitation Equipment," presents the key procedures on the building of SAS precipitation equipment, besides the evaluation of the parts' acquisition costs, and the stages of construction along with the importance of each component in the equipment. An equipment design was presented as a result of the current work that serve as a basis for consulting future works on the development of new SAS precipitation equipment.

Chapter 2, entitled "Effect of Process Conditions on the Morphological Characteristics of Particles Obtained by Supercritical Antisolvent Precipitation," presents experimental results regarding the validation of the supercritical particle formation equipment, designed and constructed by our research group, which was described in Chap. 3. It was validated using supercritical CO_2 as an antisolvent and Ibuprofen sodium salt as substance. Ethanol was used as solvent, and the effect of the operating conditions on the precipitation yield, residual organic solvent content, and particle morphology was evaluated using a split-plot experimental design and the analysis of variance (ANOVA) method.

Chapter 3, entitled "Recent Developments in Particle Formation with Supercritical Fluid Extraction of Emulsions Process for Encapsulation," discusses another variant of the SAS precipitation process called Supercritical Fluid Extraction of Emulsions (SFEE). SFEE is a strategy to process natural target compounds, because it is suitable to encapsulate poorly water-soluble drugs in an aqueous suspension, providing products with controlled particle size and increased shelf life. The rapid extraction of organic solvent favored by SFEE causes supersaturation of dispersed organic phase, which favors the precipitation of target compounds encapsulated by the polymers and surfactants. Recent reviewed data (2016 to 2018) on the feasibility of SFEE to encapsulate compounds of great interest to the food and non-food industry were provided in this chapter.

The Chap. 4, entitled "Precipitation of Particles Using Combined High Turbulence Extraction Assisted by Ultrasound and Supercritical Antisolvent Fractionation" proposes for the first time the use of combined High Turbulence Extraction Assisted by Ultrasound (HTEAU) and Supercritical Antisolvent Fractionation (SAF) of semi-defatted annatto seeds (model raw material plant) on the possibility to obtain particles with enhanced bixin and total phenolic content. The HTEAU combines two types of commercial equipments and technologies. The first is the ULTRA-TURRAX® rotor–stator technology, which produces high turbulence in the plant material bed by high extracting solvent circulation flow rate (until 2000 cm^3/min), and the second is ultrasound technology, which is recognized to improve the extraction rate by the increasing the mass transfer and possible rupture of cell wall due the formation of microcavities. The SAF produces particles with increased composition of target compounds, by the removal of solvent from the actives solution with the use of supercritical carbon dioxide as antisolvent.

Extending the new proposed designs of SAS precipitation-based process, Chap. 5, entitled "Supercritical Fluid Extraction of Emulsion Obtained by Ultrasound Emulsification Assisted by Nitrogen Hydrostatic Pressure Using Novel Biosurfactant," shows experimental results regarding the process that we named Ultrasound Emulsification Assisted by Nitrogen Hydrostatic Pressure (UEANHP), during the emulsification preparation step of the Supercritical Fluid Extraction of Emulsions (SFEE) process, one of the options of the SAS precipitation-based process. Thus in this work, first it was evaluated the influence of hydrostatic pressure levels (up to 10 bars applying nitrogen), oily phase type, and surfactant type was evaluated. In addition, the effect of an alternative biosurfactant based on saponin-rich extract obtained from Brazilian Ginseng (*Pfaffia glomerata*) roots using hot pressurized water as extracting solvent was also evaluated to further processing of this emulsion by Supercritical Fluid Extraction of Emulsions (SFEE) process, using an oily bixin-rich extract from annatto seeds (*Bixa orellana* L.) as core material (extracting solution from pressurized hot ethyl acetate extraction).

Finally, Chap. 6, entitled "Economical Effects of Supercritical Antisolvent Precipitation Process Conditions," presents simulation results regarding of the effects of several operational parameters (pressure, temperature, CO_2 flow rate, solution flow rate, injector type, and concentration of solute in the ethanol solution) during supercritical antisolvent (SAS) precipitation process on the energy consumption cost per unit of manufactured product. For this study, Ibuprofen sodium salt, as in Chap. 3, was used as a model solute and CO_2 was used as antisolvent.

Focusing on energy saving, an SAS precipitation process was simulated using the SuperPro Designer simulation platform. The effect of temperature versus concentration of ethanolic solution and pressure versus solution flow rate interactions on the energy consumption cost per unit of manufactured product was demonstrated. Thus, the present work reports a systematic energetic-economic study of the supercritical antisolvent micronization process, aiming to increase knowledge about this process and its further incorporation by the food and pharmaceutical industries.

Campinas, Brazil   Diego T. Santos
                                                                      diego_tresinari@yahoo.com.br

                                                                            Ádina L. Santana
                                                                        adina.santana@gmail.com

                                                                        M. Angela A. Meireles
                                                                        maameireles@lasefi.com

Contents

About the Authors

Diego T. Santos holds his Ph.D. in Food Engineering from University of Campinas (UNICAMP, Brazil) in 2011 and a BS degree in Chemical Engineering from University of São Paulo (USP, Brazil), 2008. Since 2011, he is working as Scientific Researcher in food engineering department at the UNICAMP. Between May 2013 and April 2014, he did postdoctoral internships at the Swiss Federal Institute of Technology (EPFL) with Prof. Dr. François Maréchal and at University of Valladolid (Spain) with Prof. Dr. Maria Jose Cocero. Besides, he did short-term research period at Dublin City University (Ireland), at the University of Chile (Chile) and at CONICET-Bahía Blanca (Argentina). He has published 79 papers in peer-reviewed journals, 9 chapters, and more than 120 works in scientific conferences. Moreover, he developed six new processes. He has participated in more than 40 research projects with both, public and private funding. He has supervised 2 Ph.D. theses, 6 M.Sc. dissertations, and 6 undergraduate research projects. His research is related to biomass valorization through the use of clean technologies Among his many activities related to the promotion of scientific development, he serves as Reviewer for 70 international journals and as Member of the editorial advisory board for 4. In addition, he also was part of the organizing committee of 7 scientific conferences. He was Lead Guest Editor for a special edition of *International Journal of Chemical Engineering* (Hindawi). He keeps scientific collaboration with several institutions: The Energy and Research Institute (TERI, Northeast Regional Centre, India), Universidad de Carabobo (Venezuela), Universidad Técnica de Machala (Equador), University of Valladolid (Spain), Federal University of Rio Grande do Norte (UFRN, Brazil), State University of Feira de Santana (Brazil), the Federal Institute of Education, Science and Technology (Capivari Campos, Brazil), and University of São Paulo (Campus Lorena, Brazil).

Ádina L. Santana holds a Ph.D. in Food Engineering from UNICAMP (2017), a M.Sc. in Chemical Engineering (2012), and a BS Degree in Food Engineering (2011) from Federal University of Pará (UFPA, Brazil). Since March 2018, she is working as a Postdoctoral Researcher Associate in the Food and Nutrition Department at the UNICAMP in the bioprocesses expertise coupled with the use of clean technologies to obtain products with enhanced quality for human health. She serves also as Reviewer for 25 international journals. In addition, she also served as Reviewer for the 5th International Conference on Agricultural and Biological Sciences—ABS 2019 (Macau, China). She has published 29 research papers in peer-reviewed journals, 8 book chapters, and 17 works in scientific conferences. She has knowledge in extraction and encapsulation of bioactive compounds with the use of supercritical fluids, pressurized liquids, and enzymes.

M. Angela A. Meireles is Director of Innovation of Natural Bioactive, Coordinator of Food Science of Coordination of Superior Level Staff Improvement (CAPES), and Professor from the Food Engineering Department at the UNICAMP (Brazil), where she began working in 1983 as Assistant Professor. In 2016, she retired from the UNICAMP where still holds the position of Professor for the Post-Graduate Program in Food Engineering and is the leading research of three technology transfer projects: (1) development of a process to obtain an extract from *Cannabis sativa*, for Entourage Lab (http://entouragelab.com/); (2) development of an integrated process to produce bioactives for cosmetic industry (https://pt-br. facebook.com/scosmeticosdobem/); and (3) assembling a supercritical fluid pilot plant to process *Algae*, for Cietec (http://www.cietec.org.br/project/bioativos/). During her time in the Academia, she taught among other courses thermodynamics, mass transfer, design, and so on at the undergraduate and graduate level. She worked closely with industries developing generally recognized as safe (GRAS) processes to obtain extracts from a variety of natural resources. After her retirement from the UNICAMP, she became Business Partner of Bioativos Naturais. She holds a Ph.D. in Chemical Engineering from Iowa State University (USA, 1982), and a M.Sc. (1979) and BS degrees (1977) in Food Engineering both from UNICAMP. She has published over 250 research papers in peer-reviewed journals and has made more than 500 presentations at scientific conferences. She has supervised 47 Ph.D. theses, 31 M.Sc. theses, and approximately 68 undergraduate research projects. Her research is in the field of the production of extracts from aromatic, medicinal, and spice plants by supercritical fluid extraction and conventional techniques, such as steam distillation and GRAS solvent extraction, including the determination of process parameters, process integration and optimization, extracts' fractionation, and techno-economical analysis of process. She has coordinated scientific exchange projects between UNICAMP and European universities in France, Germany, Holland, and Spain. Nationally, she coordinated a project (SuperNat) that involved 6 Brazilian institutions (UNICAMP, UFPA, UFRN, UEM, UFSC, IAC) and a German university (Technishe Universität Hamburg-Harburg - TUHH). In 2000–2005, she coordinated a thematic project financed by FAPESP (State of São Paulo Science Foundation) on supercritical technology applied to the processing of essential oils, vegetable oils, pigments, stevia, and other natural products. She has coordinated 4 technology transfer projects in supercritical fluid extraction from native Brazilian plants. She coordinated 2 projects in supercritical fluid chromatography to analyze petroleum in a partnership with Petrobras and to analyze food system in a partnership with the Waters Technologies of Brazil. She is

Editor-in-Chief of *The Open Food Science Journal* (https://benthamopen.com/ TOFSJ/home/), and *Food and Public Health* (http://journal.sapub.org/fph/). She is also Associate Editor of *Food Science and Technology*—Campinas (www.scielo.br/ cta). She belongs to the editorial boards of the *Journal of Supercritical Fluids*, *Journal of Food Processing Engineering* (Blackwell Publications), *Recent Patents on Engineering*, *The Open Chemical Engineering Journal* (Bentham Science Publications), *Pharmacognosy Reviews* (Pharmacognosy Networld), *Food and Bioprocess Technologies* (Springer). She was the COEIC of *Recent Patents on Engineering* from 2016 to 2017. From 1994 to 1998, she served as Associate Editor for the journals *Food Science and Technology* (Campinas) and Boletim do SBCTA (Newsletter from the Brazilian Society of Food Science and Technology). She is the editor of two books: (1) *Extracting Bioactive Compounds for Food Products: Theory and Application* (CRC Press, Boca Raton, USA) and (2) *Fundamentos de Engenharia de Alimentos* (*Food Engineering Fundamentals*—Atheneu, São Paulo, Co-editor Dr. C. G Pereira). She was Guest Editor of special issues of the *Journal of Supercritical Fluids* and *The Open Chemical Engineering Journal*.

Ademir José Petenate holds the BS degree in Computer Science (1973) and the M.Sc. degree in Statistics from the University of Campinas (UNICAMP, 1979), and the Ph.D. degree in Statistics from the Iowa State University of Science and Technology (1983). He is Professor at the UNICAMP. He coordinates IMECC/UNICAMP Extension activities and UNICAMP's Black Belt Course focused on training professionals to work in Process and Product Improvement.

Eric Keven Silva is Postdoctoral Researcher at the University of Alberta in Edmonton, Canada. He holds Ph.D. in Food Engineering from University of Campinas (2016). He performed postdoctoral internship in the School of Food Engineering at the UNICAMP (2016–2018). He has published 52 scientific articles in specialized journals, 3 chapters, and 48 scientific papers in event annals as a result of academic research activities of interaction with several co-authored collaborators. He integrated 6 research projects and received 2 academic awards. He is Reviewer of leading international journals within his research field such as *Trends in Food Science and Technology*, *Innovative Food Science and Emerging Technologies*, *Carbohydrate Polymers*, *Food Hydrocolloids*, *Food Chemistry*, *Food Research International*, *Powder Technology*, among others. He develops research in functional product engineering and process engineering with emerging technologies such as high-intensity ultrasound and supercritical carbon dioxide, focusing on non-thermal food and beverage stabilization processes; extraction processes of biopolymers, natural pigments, bioactive compounds and drugs; encapsulation processes of bioactive compounds and flavorings.

Juliana Q. Albarelli holds a BS degree in Chemical Engineering from the University of São Paulo (USP) (2003–2008), a M.Sc. degree in Chemical Engineering (2008–2009), and a Ph.D. in Chemical Engineering from the from the

University of Campinas (UNICAMP, 2009–2013) with sandwich internship at the Universidad de Valladolid (Spain) and Postdoctorate at the École Polytechnique Fédérale de Lausanne (Switzerland) (2013–2014) and at the Universidad de Valladolid (Spain, 2016). Since 2006, she has been developing research, development, and innovation activities. She acts as Reviewer for 9 international journals. She has published 40 articles in specialized journals, 2 chapters, and 46 papers in national and international scientific events. In addition, she has experience in using computational tools for evaluation and optimization of emerging processes. Additionally, she has experience in University–Company interaction projects. She participated in 34 events in Brazil and abroad, participating in the organization of 2 of them. She is currently working as Postdoctoral Researcher at the School of Food Engineering at UNICAMP. She maintains technical and scientific collaboration with several national and foreign universities such as University of São Paulo (Campus Lorena, Brazil), University of Valladolid (Spain), École Polytechnique Fédérale de Lausanne (Swiss), and the Federal Institute of Education, Science and Technology (Capivari Campos, Brazil).

Júlio C. F. Johner holds his BS Degree in Food Engineering from the Federal University of Mato Grosso (UFMT, 2010), the M.Sc. degree Food Engineering and Science from the Federal University of Rio Grande (FURG, 2013), and a Ph.D. in Food Engineering from the University of Campinas (UNICAMP, 2018). He is currently Individual Microentrepreneur at JOHNER SCIENTIFIC. Currently, he serves as Reviewer for *The Open Food Science* and the *Food Science and Technology* (Campinas). He has experience in Food Science and Technology expertise, focusing on Food Engineering.

M. Thereza M. S. Gomes is Professor of thermal and fluids at Mackenzie Presbyterian University. She holds a Ph.D. in Food Engineering from UNICAMP, in the area of Physical Separations, funded by National Council for Scientific and Technological Development (CNPq), with a sandwich doctorate from the University of Alberta, funded by the PDSE-CAPES program. She holds a master's degree in Food Engineering from UNICAMP, in the area of process engineering applied to the food industry, funded by Coordination of Superior Level Staff Improvement (CAPES). She holds a degree in Food Engineering from the University of Taubaté (2008) and has completed three scientific initiation projects funded by FAPESP.

Ricardo Abel Del Castillo Torres is Professor in the Food Engineering Department (DIA)/Faculty of Food Industries (FIA) from the National University of the Peruvian Amazon (Iquitos, Peru). He holds a Ph.D. in Food Engineering from the University of Campinas (UNICAMP, Brazil) in 2019 with the financial support from the Coordination of Superior Level Staff Improvement (CAPES, Financial code 001). In 2015, he holds his M.Sc. degree in Food Engineering at the UNICAMP with the financial support from National Council for Scientific and Technological Development (CNPq). He was the winner of the Leopold Hartman

Award, attributed from his work presented at the XXVI Brazilian Congress of Food Science and Technology (XXVI CBCTA), in the oils and fats category. He has experience in the area of Food Science and Technology, with emphasis on Food Engineering, acting on themes of environmental impact reduction through the recycling of pressurized CO_2 in pilot scale, the obtaining diversified products from pressurized fluid processes, optimization and process integration, extraction with supercritical fluids, precipitation with supercritical antisolvent, application of ultrasound technology to obtain bioactive compounds, processes for equipment design and industrial systems.

Tahmasb Hatami did his PhD in Food Engineering in UNICAMP, Brazil (2015–2018), with BSc and MSc in Chemical Engineering, Iran. During his BSc, he ranked third among Chemical Engineering students of the University of Isfahan, Iran. Moreover, in the last semester of his BSc (2007), he participated in the Iranian Olympiad of Chemical Engineering and successfully placed in the top ten. After BSc, he immediately started MSc in Chemical Engineering, Razi University, Iran. His MSc thesis was about mathematical modeling of supercritical fluid extraction for drug manufacturing. To do his MSc project perfectly, he cooperated with several international professors from McGill University (Prof. Juan H. Vera), Waterloo University (Prof. A. Elkamel), Ryerson University (Prof. A Lohi), University of Campinas in Brazil (Prof. M.A.A. Meireles), and University of Belgrade in Serbia (Prof. S. Glisic). This cooperation resulted in several publications in high-quality peer-review journals such as The Journal of Supercritical Fluid, and Fluid Phase Equilibria. After getting MSc certificate, he was employed as an academic staff member in the Department of Chemical Engineering, University of Kurdistan (UOK), Iran. His job involved teaching BSc students as well as supervising both BSc and MSc students for doing novel researches. His research covered modeling and optimization of chemical processes, supercritical fluid extraction, membrane technology, polymer processing and electrospinning, carboxymethyl cellulose production from sugarcane bagasse, and visible light photocatalyst. He was awarded the Best Professor Award (2011–2013) and the Top Researcher Award (2011–2013) in the Department of Chemical Engineering, UOK, Iran. He established the lab of petroleum engineering in the UOK, Iran, and also had cooperation in establishing heat transfer laboratory there. After that, he got admission from UNICAMP, Brazil, to do a PhD program in food engineering. After passing the required credits with the highest grade, A, Tahmasb started his PhD thesis entitled "Effects of grinding time, grinding load, and cold pressing on the aromatic compounds content of extract from fennel obtained by supercritical fluid extraction: experimental and mathematical modeling." During his PhD, he had cooperated as a visiting researcher with Dr. Ozan Ciftci in the Department of Food Science and Technology, University of Nebraska-Lincoln (UNL), USA, for six months. His project in UNL was mainly focused on experimental and modeling investigation of 1) particle formation from gas-saturated solution processes at supercritical condition, 2) drug loading to silica alcogels using supercritical technology, and 3) supercritical extraction from tomato by-product. Particularly, he modeled and optimized the aforementioned processes based on mass, momentum, and energy conservation laws. He defended his PhD thesis successfully in June 13, 2018. Now, he is a postdoc researcher in the School of Chemical Engineering, UNICAMP. Generally, he has worked with various equipment in the laboratory (such as supercritical extractor, chemical reactor, electrospinning setup, and nanoparticle production setup based on supercritical fluid), different analyzers (such as GC, HPLC, DSC, and XRD), and various software for designing, modeling, and optimization of various engineering processes (like MATLAB, COMSOL, GAMBIT, FLUENT, ASPEN, SUPERFLO, and ANSYS). Moreover, due to his teaching and supervising experiences in the university, he has very good skills in both individual and teamwork researches.

Chapter 1
A Detailed Design and Construction of a Supercritical Antisolvent Precipitation Equipment

1.1 Introduction

Some groups of compounds with polar trends do not show good solvation in the supercritical CO_2. In these cases it is common to use a cosolvent extractor which generates a final extract in solution that needs to be evaporated to obtain the pure extract. The Supercritical Anti-Solvent (SAS) technique promotes the evaporation of this solvent and allows the production of a solid extract that can be predominantly micro or nanoparticulate [1, 2].

Particle formation processes via SAS are developed in both commercial and home-made equipment. Two examples of commercial equipment are from Thar Technologies, a product that can develop the SEDS (Solution Enhanced Dispersion by Supercritical Fluids) process, a subtype of the SAS process [3]. Prior to this change in commercial equipment, some researchers who have purchased equipment for the development of conventional SAS via processes made the change in their laboratories [4]. Basically the modification occurs at the inlet of the CO_2 tubing through the insertion of a "T" connection which allows the solution containing the active substance to be micronized or encapsulated coaxially to the supercritical CO_2. Another example of equipment that develops the SAS process is the ExtrateX company that markets a multipurpose equipment that can develop the Extraction, RESS, SAS, PGSS (in option) [5].

Recently some research groups have been dedicated to assembling equipment to develop the most diverse processes involving supercritical fluid. Assembling an equipment that works with supercritical fluid without proper planning may result in poorly functional equipment. This chapter provides information on how to assemble a SAS equipment in order to indicate the steps to follow as well as the necessary parts.

© The Author(s), under exclusive license to Springer Nature Switzerland AG 2019
D. T. Santos et al., *Supercritical Antisolvent Precipitation Process*,
SpringerBriefs in Applied Sciences and Technology,
https://doi.org/10.1007/978-3-030-26998-2_1

1.2   Materials and Methods

The assembly of the unit was divided into two topics: (1) Structural part of the units; (2) Equipment and materials. In the first topic, as the equipment and materials of both units are practically the same, a general description was firstly reported, and at the end detail information about each unit was presented. Special emphasis was given in this report to the description of the operating procedures of the innovative multipurpose unit for the development of the various processes that it can operate.

1.2.1   Unit Structure

The structural part of the units was assembled with a 45 × 45 mm aluminum profile (CBA—CompanhiaBrasileira de Alumínio—SP), which has a groove on the four sides, where by means of angle metals and screws one can be able to fix one profiled bar to another. Another option would be the use of 30 × 30 mm profiling (PerfilArt Machining Bizuti, São Paulo-Brazil) that have good mechanical strength and occupy less space in the internal part of the structure [6]. This type of material facilitated the assembly, since it dispensed any type of weld or holes, requiring only fittings the parts in proper proportions. This structure (Fig. 1.1) could be assembled in the Laboratory itself without the need of a specialized workshop. Four castors, two fixed and two swivel castors were used to allow easy movement of the units.

Fig. 1.1 SAS unit structure

All components of the units were installed in the structural part through nuts and bolts called "hammer head" because one of its ends is rectangular in shape. The bolt is placed on the side of the profiled groove and when it is tightened with the nut, the bolt turns in the groove of the section, blocking in the internal part where it is fixed. With this system we can make any modification of structures as well as their components very easily.

In addition to the profiles (15 m), we used 40 angle metals, 150 bolts and nuts, 10 lids of profiles for the finishing, an aluminum plate of $100 \times 71 \times 3$ cm that was cut into a larger plate for the base of the structure and smaller plates components.

1.2.2   Equipment and Materials

- Carbon dioxide reservoir

Carbon dioxide supplied (Gama Gases Especiais, Campinas-SP) was used in cylinders with a capacity of 31 kg at a pressure of 6 MPa at 298 K, provided with a fish tube for the collection of the liquid phase of carbon dioxide, which is at the bottom, with purity (99.5%).

- In-line filters

Sintered stainless steel 5 micron ¼ inch filters (SWAGELOK–USA) were used at the inlets of the units to retain the dirt particles that remain at the bottom of the carbon dioxide reservoir, as any impurity can block the micrometer valve and the pneumatic pump heads causing their cavitation.

- Manometers

We used standard gauges (Bourdom type Record of 150 mm in diameter, São Paulo-SP), with a stainless steel body with a scale of 0–100 bar and 0–500 bar with an approximate accuracy of 0.5%. Used to measure initial pressure of the carbon dioxide cylinder and work at different points in the units. These gauges are periodically checked by the Technology Center—CT in Certified Laboratory of Metrology of UNICAMP.

- CO_2 Pneumatic Pump

A pneumatic pump (Maximator mod PP 111-VE MBR, Germany) was used in the units. This pump allows to raise the CO2 tank pressure ranging from 70 bar depending on ambient temperature to up to 600 Bar and is controlled by a valve (TESCON CORPORATION, USA) with operating capacity up to 690 psi depending on compressed air pressure enters the pump. For the operation of this pump, a dental air compressor (Schuz mod. MS 3/30, São Paulo-SP) was used, which does not use oil. After the passage through the coil, the CO_2 is fed into the pneumatic pump being pressurized. The pumpisshown in Fig. 1.2.

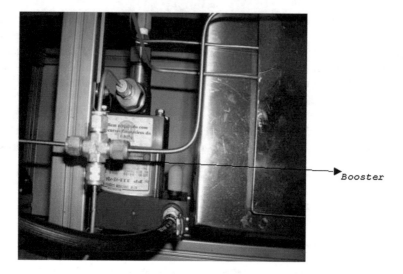

Fig. 1.2 Booster used for CO_2 pressurization in the units

In addition, a coil was mounted on the pump head according to Fig. 1.3. The constant movement of the piston can promote the heating of the pump head due to the friction between the parts which can consequently result in cavitation of the pump damaging the process. With the use of the coil in the head of the pump there is a greater guarantee of efficiency of the stage of CO_2 pumping [7].

- Thermocouples

A Type j Thermocouple with PT-100 sensor (Robert Shaw, mod. T4WN with five channels, USA) was used to measure the temperature at various points of the units. The sourcesusedwere 1/8 inchstainlesssteel.

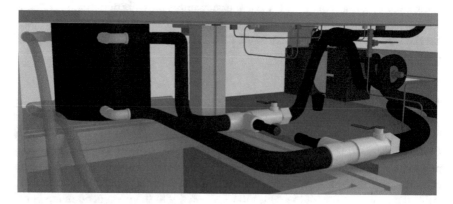

Fig. 1.3 Pump head cooling system

- Valves

Valves were used to control the flow and pressure of the units, such as:

(1) Blocking valves, exclusively to block the flow.
(2) micrometric, exclusively to control the flow. It can not be used to block, which can cause the valve to break.
(3) Pressure regulator type Back Pressure, Used exclusively for the maintenance of pressure in the line that comes before its application, it is not a flow control valve or blocker.
(4) anti-return valve used to prevent reverse flow of the fluid. The pump used contains two non-return valves which act simultaneously to promote raising the system pressure after the pump without allowing the return of the pumped fluid.
(5) safety, used to release the pressure in case the system exceeds the stipulated safety pressure.

- Blockingvalves

High pressure valves 748 bar (Autoclave Engineers, Model 10V2071 15000 psi, USA) 1/8 inch were used which allow the opening and closing of fluids to be directed by the system as shown in Fig. 1.4.

- MicrometringValves

1/8 inch micrometer valves (Autoclave Engineers, Model 10VRMM 11000PSI, USA) were used in the units. When they are opened they allow the passage of the CO_2, occurring an expansion of the CO_2 causing the temperature decrease due to the Joule-Thomson effect. In this way the valves need to be heated preventing

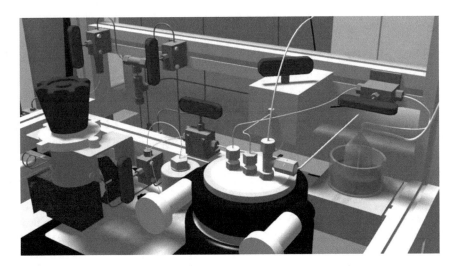

Fig. 1.4 Main valves and piping of the SAS equipment

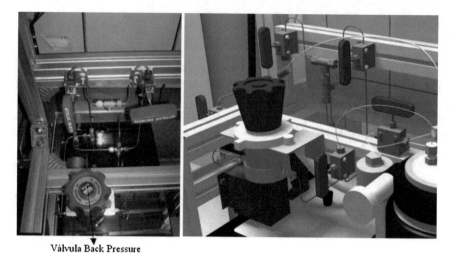

Válvula Back Pressure

Fig. 1.5 Valves, pipes and connections used in the unit

them from freezing. Heating of the micrometric valves is controlled by a temperature controller (Cole-Parmer, Dyna-Sense, Thermoregulator Control Systems Model EW-02156-40, Vernon, Illinois), which has two control outputs as shown in Fig. 1.4 [8].

- Pressure control valve (Back pressure)

We use Back Pressure Control Valves (TESCOM Model 2617-6-1-2-2-065, USA) with a capacity of up to 690 Bar. These valves were installed at the outlet of the pneumatic pumps (booster, shown in Fig. 1.5) where it operates together to control the working pressure of the units.

- Non-returnValve

Non-return valves (Swagelok series CW4BW4) were used in the units, preventing a greater pressure produced by the pump to return it. The maintenance process pressure contributes to a higher reproducibility of data collection. The equipment thus operating by recycling excess CO_2 to the line prior to CO_2 cooling allows the pump via the compressed air controller to be set at the beginning of the process together with the back pressure valve and to work until the end of the process without subsequent adjustments as long as the CO_2 supply is not interrupted.

- Safety valve

A 1/8-inch stainless steel valve (SWAGELOK, series R3A, USA) has been used which has a pressure relief ring controlled by an internal spring. The valve is supplied with a kit of springs of different colors and each color corresponds to a relief pressure. In these units, green colored springs were used, corresponding to the relief pressure of 450 Bar. Passing from this pressure the valve opens by relieving carbon dioxide, protecting the system.

- Pipes

Seamless 316 stainless steel pipes that withstand working pressures of up to 80 MPa were purchased from DETROIT Metalúrgica in diameters of 1/4, 1/8 and 1/16 inch for assembly of units (Fig. 1.3). Through these pipes we drive the flow through the units. We chose to use 1/4 pipe for CO_2 feed, 1/8 for system circulation and 1/16 for micrometer valve outlet.

- Connections

Connectors, T connectors, elbows, male connectors for thermocouples, connectors for pressure gauges and coupling, all in 316 stainless steel, were selected from the High-Seal line (Metalúrgica DETROIT, Jundiaí-SP) (Fig. 1.3). These connections are of high resistance to the pressure, temperature, allowing good sealing and easiness in the assembly and maintenance of the units.

- Thermostatic baths

To ensure that CO_2 is liquefied prior to going to the pneumatic pump, and to avoid its cavitation, MARCONI baths (mod. Ma-184, Piracicaba-SP) with external circulation were purchased to maintain cooling and condensation at—Carbon dioxide and cool the pumphead [9]. To maintain the thermal exchange of carbon dioxide prior to pressurizing, a 6 m long coil was constructed with a seamless 316 stainless steel tube having a 1 mm thick 1/8 inch wall. This serpentine was built in the laboratory by

Fig. 1.6 Thermostatic bath and coil units

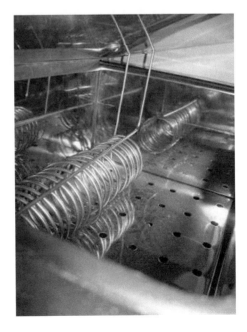

manually wrapping the tube over a piece of 75 mm PVC pipe used as a mold. In order to avoid having to construct an external heat exchanger, we install the coil inside the thermostatic bath tub, where it is submerged in a cooling solution composed of 60% water and 40% ethylene glycol, as shown in Fig. 1.6.

• Heatingbath

MARCONI heating baths (mod. MA—116/BO/E, Piracicaba, SP) were used for the previous heating of the CO_2 before its entrance into the pressure vessels. Inside this bath was placed another 1/8 in. Stainless steel coil. of diameter and 3 m in length, constructed in the same way for the thermostatic bath, in order to guarantee the same process temperature. In this way we avoid a CO_2 temperature difference that exits at $-10\,°C$ to a temperature between 40 and 60 °C that commonly has to enter the pressure vessels [10].

1.3 Results and Discussion

The constructed SAS unit can be used in public and private laboratories of research centers, universities and industries for teaching, research and development of processes and products. SAS enables precipitation unit at ambient conditions or chosen pressure and temperature or by SAS (Supercritical Fluid Anti-Solvent) among other processes.

In the processes of extraction, formation of particles, among others, generally environmentally friendly solvents are employed, such as carbon dioxide, water or ethanol, but this unit enables the use of other solvents, such as isopropanol, among others.

The unit includes fixed and mobile components, and the mobile components were built for easy maintenance and/or adaptation as the need for a new process.

The fixed components are: a pneumatic pump, a "Back Pressure regulator valve", an HPLC pump, 2 baths being, one for heating and the other for cooling some unit components, 4 manometers and 1 thermocouple for pressure and temperature measurement, respectively, at different points in the system, two different sized platforms constructed for the mobile components to be incorporated, 7 blocking valves and a micrometric valve with electric heating system to avoid freezing and line clogging due to the Joule-Thomsom effect, 1 (for the smaller pressure vessel), 2 temperature controllers (one for the micrometric valve heating system and one for the jacket), a flowmeter and a flow totalizer.

And the movable components are: 1 pressure vessel of 6.57 mL (approximately 2 cm in diameter and 4.5 cm in height) with inlet and outlet at its ends, and an ultrasonic bath; 1500 mL pressure vessel (approx. 6.5 cm in diameter and 17 cm high) with jacketed flat bottom (allows operating temperatures of -10 to 90 °C by having two coils—one for cooling fluid in the cooling bath and another for water from the heating bath) with 3 holes at its upper end which, for example, now acts as a separator in the extraction process, enabling the fractionation of the obtained

Fig. 1.7 Photo of the SAS unit

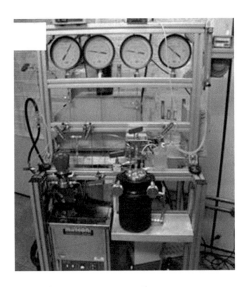

extract to function, or as a precipitation chamber in the process of particle formation via RESS and SAS, and different accessories containing small fragments of pipes previously built to enable an easy and fast rearrangement of the lines of the unit to carry out the different processes. More information on the procedures to be adopted to carry out the processes that the unit can develop, as well as flowcharts/schemes of each process are presented in the following section.

In a traditional chemical plant consisting of pipes, pumps, valves, filters, pressure vessels, etc. these elements are arranged in a predefined arrangement and any modifications desired to the plant necessitate the insertion or withdrawal of pipes, pumps, valves, etc.

In order to perform other processes which may also be carried out in the SAS unit (Fig. 1.7), such as reaction, impregnation, adsorption, particle coating using pressurized fluids, among others, it is necessary to choose which of the accessories and which of the two pressure vessels, or whether both of them, will be used. For example, for the development of extractions, only the smaller pressure vessel, together with the electric heating system and the accessories that allow the joining of the moving parts with the fixed parts, must be used.

Basically, the SAS unit can be divided into 4 parts:

Part 1—CO$_2$ feed. This part of the unit consists of pneumatic pump, a "Back Pressure regulator valve", 1 cooling bath, pipes and blocking valves. CO$_2$ is pressurized to high pressures. To this end, the solvent is first cooled by passage through an immersed coil (6 m extension) in the thermostatic bath used for cooling and thereby liquefying the solvent in order to be pumped/pressurized by the pneumatic pump [11].

Part 2—Liquid solvent feed. This part of the unit consists of HPLC pump, tubing and blocking valves. The liquid solvent (H2O, organic solvents) is pressurized to high pressures using an HPLC pump.

Part 3—Development of processes with pressurized fluid. This part of the unit is composed of 1 pressure vessels, heating bath, cooling bath, electric heating system type jacket (for the 500 mL pressure vessel), pipes, blocking valves and micrometring valve with heating system. The equipment uses the pressurized fluid [pure or a mixture (solvent + cosolvent)] for particle formation during a certain time.

Part 4—Measurement of process parameters. This part of the unit is composed of a temperature gauge (1 thermocouple) and pressure (4 gauges), temperature controllers (one for the micrometric valve heating system and another for the jacket), a flowmeter and a flow totalizer. The measurement of the parameters: process temperature and pressure, CO_2 flow and liquid solvent flow rate are continuously measured during the development of the process [12].

1.3.1 SAS Unit Operation Procedure

The following paragraphes describes the operating procedures of the SAS unit for particle formation using supercritical fluid anti-Solvent (SAS) fluids. To facilitate understanding of how the SAS unit operates a general flowchart is presented below. In this flowchart all the components that make up the unit are designed. It is noticed that part of the components are connected to others, and part not. The parts that are not connected (e.g. BL, V-5, V-6, V-7, etc.) were designed in this way, since they can be connected in different ways to enable the development of different processes, through the use of some accessories. Already the parts that are connected (example: V-1, FL, M-1, VC, etc.) will always remain as they are for the development of all processes.

In the flowcharts of the processes that the unit develops, it has been adopted that the unconnected parts will be connected to their proper places to enable the development of the desired process through the use of a thick dashed line. In addition, a list of components that will not be used during the development of each process is mentioned to avoid mistakes on the part of the operator, as well as images of how the unit should be to develop the SAS process [13].

If the aim is to: (i) form smaller sized non-encapsulated particles, prepare a homogeneous solution containing the material to be micronized in an organic solvent; (ii) forming encapsulated particles, preparing a homogeneous solution containing the material to be encapsulated together with an encapsulating material (biopolymer, cyclodextrin, etc.) in an organic solvent.

(1) Turn on the cooling bath (BR) and set it to operate at $-10\ ^\circ$C;
(2) Turn on the heating bath (BA) and program it to operate at the desired temperature;

(3) Connect the T-tubing to the central orifice of the precipitation chamber (VP-2) and its branches on the V-5 valve and the BL pump;

(4) Connect the VP-2 piping between the side (left) hole of the precipitation chamber (VP-2) and the V-6 valve;

(5) Prepare and insert a filter paper cartridge (10 μm) into the end of the VP-2 tubing (inside the precipitation chamber) to prevent a loss of drag of the particles formed in the precipitation chamber;

(6) Connect the VM/1 tubing to the cap of a sealed vial (50 or 100 mL) of known mass immersed in an ice bath;

(7) Also, connect the RT piping to the collector vial cap;

(8) Check that the valves V-1, V-2, V-5, V-6 and VM are closed;

(9) Open the CO_2 cylinder valve and valve V-1 slowly so that the pressure gauge M-1 pointer rises smoothly;

(10) Carry out pressurizing of CO_2 by opening the VC (Compressor Valve) and V-BPR (Back Pressure Regulator) valve, controlling the pressure increase (Measurement through the M-2 manometer);

(11) Seal the precipitation chamber (VP-2), connecting the thermocouple and manometer M-4 to the VP-2 vessel;

(12) When CO_2 pressure is reached, open valve V-5 to pressurize the precipitation chamber (VP-2);

(13) Turn on the controller of the micrometric valve heating system (CAVM) and program them to operate at a temperature that prevents freezing of the outlet line (this temperature depends on the CO_2 flow rate, and can vary between 343 and 393 K);

(14) When the pressure and temperature of the system are stabilized, open the valves V-6 and VM carefully until the desired flow is reached (Measurement through the rotor);

(15) Stabilize the desired flow, connect the liquid pump (BL) and program it to feed the precipitation chamber with the solution previously prepared at the desired flow rate, by the internal capillary of the coaxial system (supercritical CO_2 passes externally) (Depending on the pressure The flow rate should be a little higher than the desired flow rate—using a test tube to check if there is a correspondence between the programmed flow value and the actual flow rate, correcting it when necessary);

(16) After the pre-set particle formation time, turn off the BL pump to stop feeding the solution;

(17) For a certain time (from 20 min to 1 h), continue to feed the expansion chamber only with supercritical CO_2 at the same flow to eliminate the residual solvent in the formed particles;

(18) After the residual organic solvent has been removed, turn off the BR bath, the BA bath and the CAVM heating controller and close the valves V-5, V-1 and VC;

(19) Depressurizing the VP-2 vessel using the same flow rate used in the process of particle formation and elimination of the residual solvent so as not to cause a loss by drag of the particles formed in the precipitation chamber;

(20) Unseal the precipitation chamber (VP-2), collect, weigh and store the particles formed in freezer at the temperature equal to or less than zero for further analysis.

NOTE: When a complex mixture, such as methanolic and ethanolic extracts, is used instead of a pure substance, the process is called SAE. Therefore, for the development of the SAE process, one should follow exactly the same procedure as described.

– SAS/SAE unit flowchart [13]

Components of Fig. 1.8 are not to be considered for SAS unit mounting: V-7—Blocking Valve; V-BR-1, V-BR-2—Refrigeration Bath Valves (Control of Refrigerant Fluid Circulation).

The position of the pressure vessel connections are shown in Fig. 1.9.

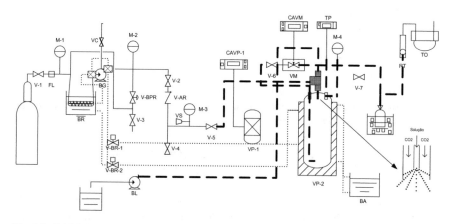

Fig. 1.8 SAS/SAE unit flowchart

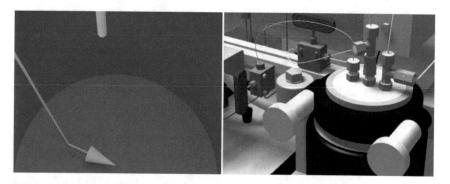

Fig. 1.9 Details of the internal and external parts of the SAS unit pressure vessel

The estimated quantities and prices of the parts and materials to be used were calculated and can be seen in Table 1.1. The cooling and heating baths with the precipitation column and the pump correspond to 49% of the assembly cost of the SAS equipment.

Table 1.1 Estimation of the cost of assembly of a SAS equipment in Brazil [6]

Product name	Qty	Price	Total (USD)
Precipitation cell system 500 mL[a]	1	1541	1541
Heatingvalve	1	157.23	157.23
Aluminumplate (m^2)	1	30.5	30.5
Aluminum profile 6 m	2.5	11.32	28.3
Angle	40	1.57	62.8
Square nut and screws	150	0.68	102
Endcap	10	0.79	7.9
Swivelcaster	4	6.92	27.68
Manometer[a]	4	341.19	1364.76
Safetyvalve	1	237.23	237.23
Temperatureindicator	1	190.25	190.25
Termocouple	1	20.35	20.35
Flowmeter	1	336.48	336.48
Flowtotalizer	1	40.88	40.88
Tubing 1/8″ (6 m)	3	89.62	268.86
FerruleandConnector 1/8″	1	549	549
CO_2filter	1	96.38	96.38
Blocking valve	7	165.17	1156.19
Micrometeringvalve	1	466.53	466.53
Back pressurevalve	1	1550.33	1550.33
Air pressureregulator	1	100.94	100.94
Air-driven CO_2 pump	1	1861.64	1861.64
Heating bath	1	1532.84	1532.84
Cooling bath	1	1691.82	1691.82
Insulation tube Ø–1/8″ (m)	1	1.83	1.83
Insulation sheets	0.6	24.45	14.67
Insulation tube Ø–22 mm	5	0.94	4.7
Silicone hose connector and adaptor	4	8	32
Silicone hose (5 m)	1	22.17	22.17
Total		13,497.26	

[a]Calculated cost

1.4 Conclusions

The designed SAS equipment in this chapter allows the development of several processes using different liquid extracts that were particulate in a solid powder composed of the bioactive materials of the original raw material. The assembly of low-cost Home-made SAS equipment can be replicated in other laboratories or even consulted by other researchers to make adaptations in home-made or commercial equipment.

Acknowledgements The authors thank the Coordination for the Improvement of Higher Education Personnel (CAPES), National Counsel of Technological and Scientific Development (CNPq), the São Paulo Research Foundation (FAPESP) and for the financial support. M.A.A. Meireles thanks CNPq for the productivity grant (302423/2015-0).

References

1. G.L. Zabot, M.A.A. Meireles, On-line process for pressurized ethanol extraction of onion peels extract and particle formation using supercritical antisolvent. J. Supercrit. Fluids **110**, 230–239 (2016). https://doi.org/10.1016/j.supflu.2015.11.024
2. V. Prosapio, I. De Marco, E. Reverchon, Supercritical antisolvent coprecipitation mechanisms. J. Supercrit. Fluids **138**, 247–258 (2018). https://doi.org/10.1016/j.supflu.2018.04.021
3. Thar, Thar Technologies (2010), http://www.thartech.com. Accessed 16 July 2010
4. G.B. Jacobson, R. Shinde, R.L. McCullough, N.J. Cheng, A. Creasman, A. Beyene, R.P. Hickerson, C. Quan, C. Turner, R.L. Kaspar, C.H. Contag, R.N. Zare, Nanoparticle formation of organic compounds with retained biological activity. J. Pharm. Sci. **99**(6), 2750–2755 (2010). https://doi.org/10.1002/jps.22035
5. Extratex (2018) Supercritical fluid extract and particle formation system. (Extratex S.A.R.L. 2018), http://www.extratex-sfi.com/equipments/sas-ress-pgss. Accessed 26 June 2018
6. J.C.F. Johner, MAdA MEIRELES, Construction of a supercritical fluid extraction (SFE) equipment: validation using annatto and fennel and extract analysis by thin layer chromatography coupled to image. Food Sci. Technol. (Camp.) **36**, 210–247 (2016)
7. G. Brunner, *Gas extraction: an introduction to fundamentals of supercritical fluids and the application to separation processes* (Springer, Steinkopff, Darmstadt; New York, 1994)
8. D.T. Santos, M.A.A. Meireles, Optimization of bioactive compounds extraction from jabuticaba (Myrciaria cauliflora) skins assisted by high pressure CO_2. Innov. Food Sci. & Emerg. Technol. **12**(3), 398–406 (2011). https://doi.org/10.1016/j.ifset.2011.02.004
9. G.L. Zabot, M.N. Moraes, A.J. Petenate, M.A.A. Meireles, Influence of the bed geometry on the kinetics of the extraction of clove bud oil with supercritical CO_2. J. Supercrit. Fluids **93**, 56–66 (2014). https://doi.org/10.1016/j.supflu.2013.10.001
10. Á.L. Santana, J.Q. Albarelli, D.T. Santos, R. Souza, N.T. Machado, M.E. Araújo, M.A.A. Meireles, Kinetic behavior, mathematical modeling, and economic evaluation of extracts obtained by supercritical fluid extraction from defatted assaí waste. Food Bioprod. Process. **107**, 25–35 (2018). https://doi.org/10.1016/j.fbp.2017.10.006
11. S. Pereda, S. Bottini, E. Brignole, Fundamentals of supercritical fluid technology, in *Supercritical fluid extraction of nutraceuticals and bioactive compounds* (CRC Press 2007), pp. 1–24. https://doi.org/10.1201/9781420006513.ch1

12. J.C.F. Johner, T. Hatami, G.L. Zabot, M.A.A. Meireles, Kinetic behavior and economic evaluation of supercritical fluid extraction of oil from pequi (Caryocar brasiliense) for various grinding times and solvent flow rates. J. Supercrit. Fluids **140**, 188–195 (2018). https://doi.org/10.1016/j.supflu.2018.06.016

13. D.T. Santos, C.L.C. Albuquerque, M.A.A. Meireles, Antioxidant dye and pigment extraction using a homemade pressurized solvent extraction system. Procedia Food Sci. **1**, 1581–1588 (2011). https://doi.org/10.1016/j.profoo.2011.09.234

Chapter 2
Effect of Process Conditions on the Morphological Characteristics of Particles Obtained by Supercritical Antisolvent Precipitation

2.1 Introduction

The use of supercritical fluids has gained importance in pharmaceutical fields. Supercritical fluid technologies for precipitating pharmaceutical substances offer several advantages over conventional ones, such as low energy requirements, low thermal and chemical degradation of products and the production of solvent-free particles with narrow particle size distributions [1]. Microparticles and nanoparticles can be formed directly from drug solutions in a single step supercritical fluid process. This could be an excellent process alternative since milling, grinding, and crushing process can lead to contamination of the product, batch variation, downstream processing difficulties, degradation of heat sensitive materials during grinding, chemical degradation due to exposure to the atmosphere, long processing times and high energy consumption [2].

Ibuprofen is a nonsteroidal anti-inflammatory drug (NSAID) widely used to treat pain, inflammation and fever [3]. The low solubility of ibuprofen in aqueous media (<1 mg mL^{-1}) limits the dissolution and absorption rates into the organism. This limitation can be overcome by the use of ibuprofen sodium salt, which can be easily dissolved in aqueous media to provide faster and greater pain relief [4]. Micronization procedures can modify particle size, porosity, and density, and the drug may be mixed with pharmaceutical excipients using small-particle technologies to maximize the delivery of the drug to the desired target during the drug administration [5].

Theoretically, a number of parameters have simultaneous effects on SAS process. Application of an experimental design approach is the most effective way to identify the most significant parameters and their interactions, and to achieve a competent result by few experimental trials [6]. Even though many parameters can influence the SAS process, some of them might not have effect on it at statistically significant levels.

© The Author(s), under exclusive license to Springer Nature Switzerland AG 2019
D. T. Santos et al., *Supercritical Antisolvent Precipitation Process*,
SpringerBriefs in Applied Sciences and Technology,
https://doi.org/10.1007/978-3-030-26998-2_2

The present work reports a systematic experimental study for micronization process using ibuprofen sodium salt as a model substance to validate the experimental equipment designed and constructed by our research group for particle formation with supercritical technology using supercritical CO_2 as an antisolvent. The SAS process was used to re-crystallize the ibuprofen sodium salt from ethanol solutions. In this context, the objective of this work was to identify which operating parameters (pressure, temperature, CO_2 flow rate, solution flow rate, injector type and concentration of ibuprofen sodium in the ethanol solution) have statistical significant effects on the precipitation yield, residual organic solvent content and particle morphology in order to obtain a better understanding of this alternative process aiming at industrial applications.

2.2 Materials and Methods

2.2.1 Materials

Ibuprofen sodium salt (BCBC9914 V, India) was purchased from Sigma-Aldrich and used as a model substance in the precipitation experiments. Ethanol (Dinâmica®, 52,990, Diadema, Brazil), with a minimum purity of 99.5%, was used to prepare the ibuprofen sodium solutions. CO_2 (99% purity, Gama Gases Especiais, Campinas, Brazil) was used as the antisolvent in the SAS process.

2.2.2 Experimental Procedure

A schematic diagram of the constructed experimental setup to perform the SAS precipitation experiments on a laboratory scale is shown in Fig. 2.1. The procedure was performed as follows: The CO_2 from the cylinder was cooled to -10 °C using a thermostatic bath (Marconi, MA-184, Piracicaba, Brazil) to ensure that liquid CO_2 is being pumped by the air driven pump (Maximator, M111 CO_2, Germany) in a 500 mL AISI 316 stainless steel precipitation vessel with a 6.8 cm inner diameter. The precipitation vessel was fitted with an electric heating jacket and an AISI316 stain less steel porous filter (screen size of 2 μm) fixed at the bottom of the vessel, which was used to collect the precipitated particles.

Once the desired conditions of pressure, temperature and CO_2 flow rate were achieved and remained stable, the ethanolic solution, which contains ibuprofen sodium, was introduced into the vessel by a high pressure pump (Jasco, PU-2080, Japan), which allows a maximum working solution flow rate of 10 mL min^{-1}. A volume of 43 mL was injected into the precipitation vessel, and 10 mL of pure ethanol was then pumped to clean the tubes. Depending on the solution flow rate used, the time allowed for precipitation was 43 or 86 min.

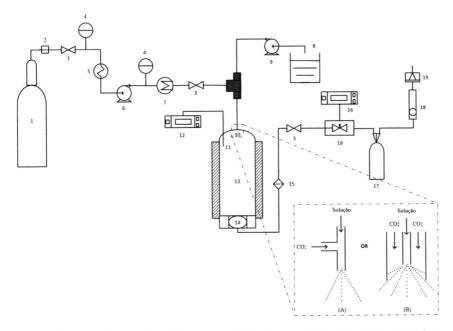

Fig. 2.1 Schematic diagram of the SAS apparatus. 1 CO_2 Cylinder; 2 CO_2 Filter; 3 Blocking Valves; 4 Manometers; 5 Thermostatic bath; 6 Air driven pump; 7 Heating bath; 8 Solution (solute/solvent) reservoir; 9 High pressure pump; 10 Injector; 11 Thermocouple; 12 Temperature controllers; 13 Precipitation vessel (Operation conditions: 10 and 12 MPa; 313 and 323 K); 14 Filter; 15 Line filter; 16 Micrometric valve with a heating system; 17 Glass flask; 18 Glass float rotameter; 19 Flow totalizer

In this work, two different injectors were used to mix CO_2 and the solution at the inlet of the precipitation vessel: (i) a home-made coaxial nozzle, which consists of a stainless steel tube with an inner diameter of 1/16 in (i.d. 177.8 mm) for the solution, placed inside a 1/8 in stainless steel tube for the CO_2; (ii) a commercial 1/8 in stainless steel T-fitting. The injectors were placed at the top of the precipitation vessel. Figure 2.1 also shows schematic diagrams of the two injectors.

When the ethanolic solution and CO_2 were mixed, the ethanol was quickly solubilized by the supercritical CO_2, and this fluid mixture (CO_2 plus ethanol) exited the vessel and flowed to a glass flask (100 mL) connected to a micrometric valve. This valve was maintained at 393 K to avoid the freezing and blockage of the outlet caused by the Joule–Thompson effect of the expanding CO_2. Ethanol was deposited in the glass flask, and the gaseous CO_2 was discharged to the atmosphere. The temperature and pressure were measured with instruments directly connected to the precipitation vessel with accuracies of ± 2 K and ± 0.2 MPa, respectively. The CO_2 flow rate was measured using a glass float rotameter (ABB, 16/286A/2, Warminster, USA) coupled with a flow totalizer (LAO, G0,6, Osasco, Brazil).

After the injection of pure ethanol, the high pressure pump was stopped and only CO_2 was pumped, using a minimum of 300 g of CO_2 to ensure that all remaining traces of ethanol present in the precipitation vessel were removed before depressurization. Subsequently, the air driven pump was stopped, the precipitation vessel was slowly depressurized to atmospheric pressure by manually opening the valves and the particles were retained inside the precipitation vessel by a porous filter fixed at the bottom of the vessel and another placed at the vessel outlet (AISI316 stain less steel porous line filter (Hoke, 6321G2Y, United States), porosity of 2 μm). The particles were carefully collected and stored at ambient temperature in a glass desiccator protected from light until subsequent analysis.

Sample Preparation

If the goal is: (i) Micronization, solubilise the solute in an organic solvent at the concentrations desired; (ii) Encapsulation: solubilise the solute and the wall material (e.g., biopolymer, ciclodextrin) in an organic solvent at the concentrations desired. Make sure there is no precipitation. If necessary keep in ultrasonic treatment and/or use the centrifuge for eliminating the precipitate.

1. Turn on the thermostatic bath and setup the temperature of -10 °C. Wait the temperature achieves -10 °C to start;
2. Turn on the heating bath and setup temperature of work;
3. Turn on the high pressure pump and setup the solution flow rate of work;
4. Turn on the temperature controller and set up the temperature of work;
5. Prepare the precipitation vessel by placing the ring in the bottom and upper parts of the vessel. Insert the AISI 316 stainless steel porous filter at the bottom of the vessel. Close the bottom and upper parts;
6. Connect the inlet and outlet tubes to the precipitation vessel;
7. Connect the tubes to the glass flask;
8. Check if all blocking valves are closed;
9. Open the CO_2 cylinder;
10. Pressurize the precipitation vessel using the air driven pump until the pressure of work;
11. Check for leaking in the system;
12. Turn on the heating system of the micrometric valve and setup a temperature between 70 and 120 °C, depending on CO_2 flow rate used, to avoid freezing;
13. When the pressure and temperature of work remain stable, open the outlet valve until the desired CO_2 flow rate be achieved e remain stable (Read from the rotameter);
14. Run the high pressure pump during the desired time;
15. Stop the high pressure pump;
16. Keep CO_2 flow rate during an adequate time to eliminate residual solvent;
17. Turn off the temperature controller, thermostatic and heating bath and close all valves;
18. Depressurize slowly the system;
19. When no more pressure inside the vessel, disconnect all connections, remove and open the precipitation vessel and collect the formed particles.

A conventional solvent evaporation (CSE) process of ibuprofen sodium process was also used as a reference process to compare it with the SAS process. Two ethanolic solutions were prepared at 0.02 and 0.04 g mL^{-1} and the ethanol was evaporated using a rotary evaporator (Tecnal, TE-211, Piracicaba, Brazil) with a vacuum control of 200 mmHg and thermostatic bath at 313 K. The ibuprofen sodium precipitates were collected from the precipitation vessel (50 mL glass flask) and stored at ambient temperature in a glass desiccator protected from light. In previous studies the CSE method has been used to produce microparticles of various solids (e.g. phospholipids complex of puerarin; felopidine).

2.2.3 Experiment Design and Statistical Analysis

Fractional factorial designs are widely used in industrial experiments. Completely Randomized Design (CRD) is usually used in this situation where the treatments are completely randomized to the experimental units. In this work, an experiment with 5 factors, each at two levels, was conducted. However, it is very difficult to change the levels for the factor "type of injector" and a CRD would eventually require a modification of the apparatus after each experimental run. Due to time constraint and to avoid leaks, the factor "type of injector" was considered as a hard-to-change parameter. In this case, a fractional factorial split-plot design represents a practical design option and the experimental runs were done accordingly to this method. A discussion about this experimental design technique can be found in Box, Hunter and Hunter [7].

In the split-plot design, the hard-to-change factor is called whole-plot and the easy-to-change factors sub-plots. First the whole-plot factor is randomized and in the sequence the sub-plots are randomized within the whole plot. While holding the level of the factor "type of injector" fixed, all of the level combinations of the remaining factors are randomized in a random order was run.

The injection type (T-fitting and coaxial nozzle) was applied to the whole-plots with two replications and a 2^{5-2} fractional factorial while considering the temperature (313 and 323 K), pressure (10 and 12 MPa), concentration of ethanolic solutions (0.02 and 0.04 g mL^{-1}), CO_2 flow rate (500 and 800 g h^{-1}) and solution flow rate (0.5 e 1.0 mL min^{-1}) applied to the sub-plots, which totaled 32 experimental units. Treatments were deemed to be statistically significant for p-value <0.05 (95% confidence level). Statistical analysis was conducted with MINITAB Statistical Software (Minitab Inc., State College, Pennsylvania). The data from different experiments were compared for statistical significance by analysis of variance (ANOVA).

2.2.4   Analysis and Characterization

2.2.4.1   Determination of Residual Organic Solvent

Gas chromatography (GC) was used to determine the residual amount of ethanol in the particles. The residual solvent was analyzed on a Shimadzu gas chromatograph (GC-17-A, Kyoto, Japan) equipped with flame ionization detection (FID) system. Approximately 30 mg of sample was dissolved in 1 mL of toluene with the aid of an ultrasonic bath (Unique, Max Clean 1400, 40 Hz, Indaiatuba, Brazil). Sample solutions (1 μL) were introduced by direct injection on a Zebron ZB-5 capillary column from Phenomenex (30 m × 0.25 mm and 0.25 μm). The other conditions were: injection temperature of 493 K; detector temperature of 513 K; helium flow rate of 28 mL min^{-1} and split ratio of 1:20. Helium served as the carrier gas, and the analysis was performed using an oven temperature of 313 K with a ramp of 20 K min^{-1} until a temperature of 453 K was reached. The data were quantified using a calibration curve that was constructed by measuring different known concentrations of ethanol in toluene.

The chromatography profile of the standard solution containing ethanol and toluene showed an ethanol peak at a retention time of 2.04 min. The resultant linear relationship between peak area (y) and ethanol concentration (x) was y = 32,719 x − 1030, $R^2 = 0.9998$.

2.2.4.2   Determination of Morphology

The morphology of ibuprofen sodium particles was examined by scanning electron microscopy (SEM; LEO Electron Microscopy/Oxford, Leo 440i, Cambridge, England) with an energy dispersive X-ray analyzer (LEO Electron Microscopy/Oxford, 6070, Cambridge, England). The samples were coated with a thin layer of gold in a Polaron sputter coater (VG Microtech, SC7620, Uckfield, England) and examined using a SEM at 20 kV accelerating voltage and 100 pA beam current. A coarse measurement of the particle size was done over the micrographs from the SEM analysis.

2.3   Results and Discussion

The effect of the temperature, pressure, CO_2 flow rate, solution flow rate, concentration of the ethanolic solution and type of injector on the characteristics of ibuprofen sodium particles obtained via SAS was investigated. The range of the experimental conditions adopted in this work was based on information from the literature [8] and preliminary tests. Table 2.1 shows the experimental conditions used for the SAS experiments randomized by the split-plot experimental design.

Table 2.1 Experimental conditions from split-plot design and results obtained in each experiment

Exp.	T (K)	P (MPa)	CO_2 flow rate (g h^{-1})	Solution flow rate (mL min^{-1})	Concentration of ethanolic solution (g mL^{-1})	Injector	Precipitation yield (%)	Residual solvent content (mg kg^{-1})	Morphology
12	313	10	500	0.5	0.02	Coaxial	26.77	10.1 ± 0.5	Flake and sheet
2	313	10	500	0.5	0.04	T-fitting	41.68	9 ± 2	Flake
31	313	10	500	1	0.02	T-fitting	27.9	5.1 ± 0.5	Sheet
19	313	10	500	1	0.04	Coaxial	14.72	16.3 ± 0.9	Flake
6	313	10	800	0.5	0.02	T-fitting	69.99	4.7 ± 0.3	Sheet
15	313	10	800	0.5	0.04	Coaxial	25.44	54 ± 1	Flake, needle and sheet
17	313	10	800	1	0.02	Coaxial	35.82	7.3 ± 0.1	Flake, needle and sheet
32	313	10	800	1	0.04	T-fitting	66.82	9 ± 2	Flake and needle
30	313	12	500	0.5	0.02	T-fitting	21.48	5 ± 1	Flake and needle
23	313	12	500	0.5	0.04	Coaxial	16.92	13 ± 1	Flake and sheet
16	313	12	500	1	0.02	Coaxial	21.33	13 ± 1	Flake and needle
3	313	12	500	1	0.04	T-fitting	25.75	6.0 ± 0.1	Flake and sheet
21	313	12	800	0.5	0.02	Coaxial	41.86	15 ± 2	Needle and sheet
26	313	12	800	0.5	0.04	T-fitting	41.13	9.3 ± 0.2	Flake
1	313	12	800	1	0.02	T-fitting	38.95	44 ± 6	Flake and needle
10	313	12	800	1	0.04	Coaxial	35.34	6 ± 1	Flake and needle

(continued)

Table 2.1 (continued)

Exp.	T (K)	P (MPa)	CO_2 flow rate (g h^{-1})	Solution flow rate (mL min^{-1})	Concentration of ethanolic solution (g mL^{-1})	Injector	Precipitation yield (%)	Residual solvent content (mg kg^{-1})	Morphology
22	323	10	500	0.5	0.02	Coaxial	24.15	4.7 ± 0.1	Flake and needle
29	323	10	500	0.5	0.04	T-fitting	30.48	6.1 ± 0.6	Flake
4	323	10	500	1	0.02	T-fitting	23.45	9 ± 1	Flake and needle
14	323	10	500	1	0.04	Coaxial	37.07	41 ± 1	Flake
25	323	10	800	0.5	0.02	T-fitting	30.36	6.4 ± 0.6	Sheet
18	323	10	800	0.5	0.04	Coaxial	27.32	16.6 ± 0.3	Flake
11	323	10	800	1	0.02	Coaxial	20.68	37 ± 2	Flake and sheet
7	323	10	800	1	0.04	T-fitting	35.88	7.0 ± 0.1	Flake
8	323	12	500	0.5	0.02	T-fitting	19	6.7 ± 0.3	Flake, needle and sheet
9	323	12	500	0.5	0.04	Coaxial	25.03	5.1 ± 0.4	Flake and needle
24	323	12	500	1	0.02	Coaxial	29.41	3.7 ± 0.7	Flake and needle
27	323	12	500	1	0.04	T-fitting	25.86	5.3 ± 0.2	Flake and needle
13	323	12	800	0.5	0.02	Coaxial	25.99	21 ± 2	Flake
5	323	12	800	0.5	0.04	T-fitting	31.98	4.3 ± 0.3	Flake and needle
28	323	12	800	1	0.02	T-fitting	72.58	9.9 ± 0.2	Flake and needle
20	323	12	800	1	0.04	Coaxial	53.57	27 ± 2	Flake

The experiments were performed using typical conditions for SAS. The pressure and temperature values were chosen in order to operate the process above the critical point of the ethanol and CO_2 mixture, which was located at approximately 8.5 MPa at 313 K [9]. Martín, Scholle, Mattea, Meterc and Cocero [8] did not obtain a dry ibuprofen sodium powder at subcritical conditions because the formation of particles is controlled by the fluid mechanics and the kinetics of evaporation of the solvent under these conditions, and the time available inside the precipitation vessel was not sufficiently long to completely evaporate the solvent in their study. The operation above or below the mixture critical point results in different particle formation mechanisms in the SAS. Su, Lo and Lien [10] observed that when operating above the critical point, no droplet formation and fast mass transfer after solution injection favored the production of smaller fluticasone propionate particles.

The choice of the solution and CO_2 flow rates was related to the difficulties in producing a dry powder. Preliminary tests showed that obtaining a dry powder was not possible at higher solution flow rates and lower CO_2 flow rates. Under these conditions, the mass of CO_2 used was not sufficient to completely eliminate the organic solvent and precipitate the ibuprofen sodium. On the other hand, CO_2 flow rates higher than 1000 g h^{-1} resulted in a higher ibuprofen sodium loss in the downstream separator. Thus, moderate CO_2 and solution flow rates of 500–800 g h^{-1} and 0.5–1.0 mL min^{-1}, respectively, were adopted in this study.

2.3.1 Influence of the Operating Conditions on the Precipitation Yield

Table 2.1 presents the precipitation yields obtained from each experiment. The precipitation yield ranged from 14.72 to 72.58% depending on the operating conditions used in the SAS process. A precipitation yield of 100% was obtained using the conventional solvent evaporation (CSE) method. The lower yield of the SAS experiments can be attributed to the loss of individual particles through the filters. The individual particle size was approximately 1 µm, which is smaller than the filter pore size of 2 µm. Thus, only agglomerates could be retained. Furthermore, slower precipitation kinetics can lead to a loss of ibuprofen sodium in the downstream separator. These reasons were also responsible for the low yield achieved by Visentin's group study [11] for the precipitation and encapsulation of rosemary antioxidants by SAS. Using a filter with a smaller pore size may result in a greater precipitation yield. However, such a filter may also become clogged more easily. A satisfactory yield above 85% was obtained by Su, Lo and Lien [10] in the micronization of fluticasone propionate using dichloromethane as the solvent, and a lower yield, of about 70% was obtained when particle adhesion on the surface of the spiral jet mill occurred.

SAS formulations of oxeglitazar with various solubilizing excipients were precipitated using six different organic solvents by Majerik and his group [12]. The obtained precipitation yield varied between 28 and 91%. The authors observed that precipitation from non-chlorinated solvents resulted in low precipitation yield suggesting that these solvents act as cosolvent and increase the solubility of processed pharmaceutical ingredients in the supercritical phase. This phenomenon was considered by the authors as the main source of loss in precipitation yield using supercritical fluid antisolvents and must be considered in this work, since ethanol, a non-chlorinated solvent was used.

The data from different experiments were compared for statistical significance by analysis of variance (ANOVA). The p-values for the precipitation yield are given in Table 2.2. According to these values, only CO_2 flow rate significantly influenced the precipitation yield (p-value <0.05). Table 2.3 shows that the precipitation of ibuprofen sodium salt at CO_2 flow rate of 800 g h^{-1} gave a significantly higher precipitation yield compared to precipitation at 500 g h^{-1}. For the interpretation of this result, it must be taken into account that an increase in the CO_2 flow rate leads to a lower contact time between CO_2 and the solute. As a result, a higher precipitation yield was achieved by means of the lower amount of dissolved solute in the supercritical CO_2 and of the less dragged solute. Moreover, the higher CO_2 flow rate enhanced the turbulence, due to the increase in the Reynolds number [13], which resulted in a better mixing between the solvents contributing to the precipitation process. Imsanguan, Pongamphai, Douglas, Teppaitoon and Douglas [14] observed the same influence of CO_2 flow rate in the precipitation yield of andrographolide from A. paniculata extract using SAS process. The authors reported that CO_2 flow rate affects the kinetic energy of supercritical CO_2 and the composition of the fluid phase. It is noteworthy that a maximum precipitation yield is intended and CO_2 flow rate of 800 g h^{-1} achieved optimum results, and preliminary studies showed that further increasing leads to a higher loss of the ibuprofen sodium in the downstream separator.

Table 2.2 P-values obtained statistically for precipitation yield and residual solvent content

Parameter	Precipitation yield	Residual solvent content
Injector	0.070	0.255
Temperature	0.599	0.834
Pressure	0.867	0.558
CO_2 flow rate	0.009	0.162
Solution flow rate	0.386	0.483
Concentration of ethanolic solution	0.943	0.693

Table 2.3 Influence of statistically significant parameters on precipitation yield (%)

CO_2 flow rate (g h^{-1})	Precipitation yield (%)[a]
500	25.69 ± 6.90
800	40.86 ± 16.36

[a]Mean ± standard deviation

As can be observed at Table 2.3, the variability in precipitation yield was observed with higher CO_2 flow rate. The increase of the deviations can be related to an unstable deposition of the particles on the filter allowing the exit of them.

2.3.2 Influence of Operating Conditions on Residual Organic Solvent Content in the Micronized Particles

The results presented in Table 2.1 show a low residual organic solvent content in the ibuprofen sodium precipitated by SAS. The residual ethanol content in the ibuprofen sodium particles was below 55 mg kg^{-1} in all SAS experiments, while the ibuprofen sodium particles precipitated by CSE was 2.5 times higher, approximately 140 mg kg^{-1}. These concentrations are lower than the suggested value of the Internal Conference on Harmonization (ICH) guideline Q3C (Impurities: Guideline for Residual Solvents), which for ethanol is 5000 mg kg^{-1} [15]. Also, the ethanol concentrations in the particles are significantly different, indicating that the SAS process has technical advantages to obtain solvent-free products. Majerik, Charbit, Badens, Horváth, Szokonya, Bosc and Teillaud [12] also demonstrated that SAS reduces residual solvent content more efficiently than CSE. The authors observed that the residual dicloromethane content and in the solid dispersions of oxeglitazar in PVP K17 (polyvinilpyrrolidone) obtained by CSE was 1.4 higher than the one obtained by SAS, and the solid dispersions of oxeglitazar in poloxamer 407 (polyoxyethylene-polyoxypropylene block copolymer) was 1.6 higher.

Similar results were obtained by Adami, Reverchon, Järvenpää and Huopalahti [16]. At certain experimental conditions, it was reported the production of micronized nalmefene HCl via SAS with a residual organic solvent content of 2 mg kg^{-1}. In a study by Kim, Lee, Park, Woo and Hwang [17], GC analysis revealed that the residual solvent (dichloromethane) content in the precipitated cilostazol via SAS was below 50 mg kg^{-1}, indicating that the solvent removal during SAS, depends not only on volatility of the solvent but also on other factors, such as the operating conditions and the SAS apparatus.

Depending on the operating conditions, the residual organic solvent could be reduced below to 4 mg kg^{-1}. The statistical analysis showed that all the evaluated operating parameters did not influence the residual solvent content in the studied range (Table 2.2). On the other hand, an interesting behavior was observed: the experiments that resulted in the best precipitation yield also showed low residual solvent content (using a T-fitting; at high CO_2 flow rate).

2.3.3 Influence of Operating Conditions on Morphology

The effect of different process parameters on the morphology of ibuprofen sodium particles obtained with the SAS process was studied. Figure 2.2 shows a scanning electron microscopy (SEM) picture of unprocessed ibuprofen sodium. The unprocessed ibuprofen sodium consisted of typical agglomerated sheet-like particles [8] with particle sizes of approximately 30 μm. The morphology obtained in each experiment is presented in Table 2.1.

Ibuprofen sodium particles were successfully micronized with a SAS process. Figure 2.3 shows the SEM pictures of the particles obtained from various experiments listed in Table 2.1. The particles of ibuprofen have a trend to develop agglomerates after the SAS process, as shown in Fig. 2.3. A coarse measurement of the agglomerates was done from the micrographs of the SEM analysis. For example, Exp. 20 shows agglomerate of ≈5 μm of diameter and Exp. 8 show a bulk set of agglomerates. The measurement was not possible for all the micrographs due to the lack of visible limit between agglomerate.

The particle morphology depended on the experimental conditions and included sheet-(Ex. Experiment 20), flake-(Ex. Experiment 6) and needle- (Ex. Experiment 16) like morphologies (Fig. 2.3). According to Reverchon, Caputo and De Marco [18], the particle morphology can be affected by many process parameters, such as the temperature, pressure, solution concentration and antisolvent to solution flow rate ratio. Moreover, the process arrangement and the apparatus can also heavily influence the product properties. In most experiments, needle-like particles of micron- to submicron size were obtained, but these particles were highly agglomerated and formed star-shaped aggregates. Nevertheless, appropriate operating conditions have been shown to generate particles that were approximately 1 μm in size. Bakhbakhi, Alfadul and Ajbar [19] obtained submicron-sized ibuprofen sodium particles with a needle-like morphology by SAS, which corroborates our findings.

(a)   **(b)**

Fig. 2.2 SEM image of unprocessed ibuprofen sodium particles with magnification of 1000 (**a**) and 3000 (**b**)

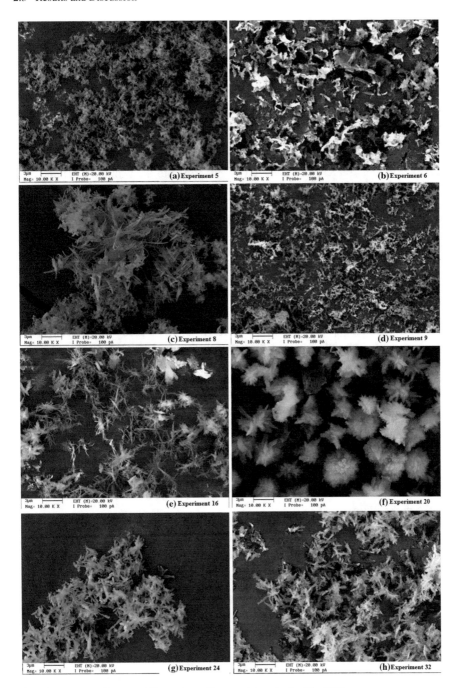

Fig. 2.3   a–h SEM images of ibuprofen sodium particles obtained by SAS process

The different morphology among agglomerates from SAS and CSE process maybe trigger for a different molecular structure of the ibuprofen. Previous studies showed that ibuprofen is a polymorphic drug [8] indicating that different conditions of pressure, temperature and solvent can induce a more stable molecular structure. Therefore, the produced agglomerate can present different physico-chemical properties. These characterizations are beyond the goal of this study, but future research is encouraged to understand the influence of the different molecular structure of the ibuprofen on SAS.

The influence of the CO_2 and solution flow rate can be explained by Martín, Scholle, Mattea, Meterc and Cocero [8]. According to the authors, when the ratio CO_2/solution flow rate is reduced, the solute solubility is increased in the fluid due to the cosolvent effect of the organic solvent, which promotes slower nucleation kinetics that produces fewer particles with bigger sizes.

The cosolvent effect is well known in supercritical extraction processing and has been used to improve the solubility of otherwise difficult to solubilize compounds by altering the polarity of CO_2 [20]. On the other hand, the results of this work showed that higher CO_2/solution flow ratios resulted in particles with a sheet morphology that were significantly less agglomerated but bigger than the flake- and needle- like particles. This effect is particularly noticeable in the results of experiments 6, 21 and 25, which operated at CO_2/solution flow ratio of 34 and produced large sheet-like particles with a particle size of approximately 5 μm. Though not predominant, a sheet-like morphology was produced in experiments 8, 9, 12 and 29 (operated at CO_2/solution flow ratio of 21), as demonstrated in Fig. 2.3c. Experiments 5, 10 and 16 produced smaller particles with a needle-like morphology.

Under supercritical conditions, the density of CO_2 is influenced by pressure and temperature and plays an important role for mass transfer between organic solvents and CO_2 during particle formation [17]. The SEM images indicated that the density of CO_2 can alter the particle size because it alters the mass-transfer characteristics of the process. Smaller particles were obtained using CO_2 at a density of 587 kg m^{-3}. Density data for CO_2 was taken from NIST [21] according to the pressure and temperature used.

The ability to alter the sizes and morphology of drug particles is important to their formulation and administration. Pathak, Meziani, Desai and Sun [22] applied Rapid Expansion of a Supercritical Solution into a Liquid Solvent (RESOLV) to produce exclusively nanoscale ibuprofen particles in aqueous suspension using polymers as stabilization agent. The authors concluded that the experimental conditions and the selection of stabilization agent in RESOLV are used to alter the sizes and morphology of the nanosized drug particles.

Ibuprofen sodium was also micronized using a conventional process, which was used as a reference process to compare with the SAS process. The SEM images of ibuprofen sodium particles obtained by conventional solvent extraction (CSE) are presented in Fig. 2.4. The ibuprofen sodium morphology did not change after precipitation; sheet-like particles were produced using CSE. However, the ibuprofen sodium particles obtained by CSE were 3 times bigger than those of the unprocessed ibuprofen sodium. The CSE method involves multi-stage processing to reduce the

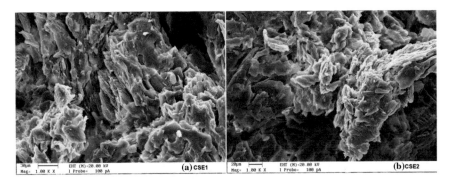

Fig. 2.4 SEM images of ibuprofen sodium particles obtained by CSE using concentrations of ethanolic solution of **a** 0.02 e **b** 0.04 g mL^{-1}

particle size and is always time-consuming. The solid obtained by Li, Yang, Chen, Chen and Chan [23] using CSE was collected and desiccated in a vacuum oven at 323 K, then pulverized and finally passed through a 180-mm sieve and Won, Kim, Lee, Park and Hwang [24] produced solid dispersions of felodipine by CSE and pulverized the product using a mill to reduce the particle size.

2.3.4 Selecting Appropriate Conditions

In the SAS equipment used in this work, our findings indicated that to produce ibrupofen sodium salt particles with a low residual solvent and high precipitation yield, the process must operate at CO_2 flow rate of 800 g h^{-1}, solution flow rate of 1 mL min^{-1} and concentration of ethanolic solution of 0.04 g mL^{-1}, which experiment 32 yields satisfactory results for all these response variables. However, flake and needle-like particles were obtained at this condition (Fig. 2.3h).

The particle morphology has been shown to affect the flow properties and tableting performance. Small particles with an adequate morphology could improve the bioavailability. Historically, compounds that exhibit a needle-like crystalline shape have showed poor flow properties, and ibuprofen is not any exception to this observation. Sheet-like particles show improved flow and tableting properties compared to needle-like particles [25]. Homogeneous sheet-like micronized particles were produced at experiment 6 (Fig. 2.3b), with also low residual solvent content (4.7 mg kg^{-1}) and high precipitation yield (70%).

2.4 Conclusions

Ibuprofen sodium salt was successfully micronized by SAS using experimental equipment designed and constructed by our research group. When the ethanolic solution comes into contact with supercritical CO_2, both are expanded, inducing phase separation and supersaturation of the ibuprofen sodium in supercritical CO_2, leading to its nucleation and precipitation. The influence of the operating parameter during ibuprofen sodium salt SAS micronization was investigated in deep by means of experimental design and proper statistical analysis. The CO_2 flow rate influenced the precipitation yield at statistically significant levels meanwhile none operating parameters did influence the residual solvent content in the micronized particles. Selecting appropriate process conditions (using the T-fitting at a temperature of 313 K, pressure of 10 MPa, CO_2 flow rate of 800 g h^{-1}, solution flow rate of 0.5 mL min^{-1} and ethanol solution concentration of 0.02 g mL^{-1}) has shown to facilitate the production of homogeneous sheet-like microparticles of ibuprofen particles, the best for tableting purposes, with low residual solvent (4.7 mg kg^{-1}) and high precipitation yield (70%).

Acknowledgements The authors are grateful to CNPq (470916/2012-5) and FAPESP (2012/10685-8) for their financial support. M. Thereza M. G. Rosa and Eric Keven Silva thanks CNPq (140641/2011-4 and 140275/2014-2) for the Ph.D. assistantship. Diego T. Santos thanks the FAPESP (10/16485-5; 12/19304-7) and CAPES for the postdoctoral fellowships. M. Angela A. Meireles thanks CNPq for a productivity grant (301301/2010-7). The authors also thank Moyses N. Moraes for his assistance with the statistical analyses.

References

1. Z. Knez, E. Weidner, Particles formation and particle design using supercritical fluids. Curr. Opin. Solid State Mater. Sci. **7**, 353–361 (2003)
2. S. Dalziel, G. Foggin, W. Ford, H. Gommeren, High pressure media milling system and process of forming particles, Patent US 20050258288 A1, Google Patents (2004)
3. P.J. Linstrom, W. Mallard, NIST Chemistry, National Institute of Standards and Technology. Gaithersburg (2003)
4. P. Gruber, M. Reher, Dosage form of sodium ibuprofen, Patent US 20040102522 A1, Google Patents (2004)
5. T.L. Rogers, K.P. Johnston, R.O. Williams 3rd, Solution-based particle formation of pharmaceutical powders by supercritical or compressed fluid CO2 and cryogenic spray-freezing technologies. Drug Dev. Ind. Pharm. **27**, 1003–1015 (2001)
6. K.M. Sharif, M.M. Rahman, J. Azmir, A. Mohamed, M.H.A. Jahurul, F. Sahena, I.S.M. Zaidul, Experimental design of supercritical fluid extraction—A review. J. Food Eng. **124**, 105–116 (2014)
7. G.E. Box, J.S. Hunter, W.G. Hunter, *Statistics for experimenters: design, innovation, and discovery*, 2nd edn. (Wiley, New York, 2005)
8. Á. Martín, K. Scholle, F. Mattea, D. Meterc, M.J. Cocero, Production of Polymorphs of Ibuprofen Sodium by Supercritical Antisolvent (SAS) Precipitation. Cryst. Growth Des. **9**, 2504–2511 (2009)

9. C.J. Chang, K.-L. Chiu, C.-Y. Day, A new apparatus for the determination of P–x–y diagrams and Henry's constants in high pressure alcohols with critical carbon dioxide. J. Supercrit. Fluids **12**, 223–237 (1998)

10. C.S. Su, W.S. Lo, L.H. Lien, Micronization of fluticasone propionate using supercritical antisolvent (SAS) process. Chem. Eng. Technol. **34**, 535–541 (2011)

11. A. Visentin, S. Rodríguez-Rojo, A. Navarrete, D. Maestri, M.J. Cocero, Precipitation and encapsulation of rosemary antioxidants by supercritical antisolvent process. J. Food Eng. **109**, 9–15 (2012)

12. V. Majerik, G. Charbit, E. Badens, G. Horváth, L. Szokonya, N. Bosc, E. Teillaud, Bioavailability enhancement of an active substance by supercritical antisolvent precipitation. J. Supercrit. Fluids **40**, 101–110 (2007)

13. X. Sui, W. Wei, L. Yang, Y. Zu, C. Zhao, L. Zhang, F. Yang, Z. Zhang, Preparation, characterization and in vivo assessment of the bioavailability of glycyrrhizic acid microparticles by supercritical anti-solvent process. Int. J. Pharm. **423**, 471–479 (2012)

14. P. Imsanguan, S. Pongamphai, S. Douglas, W. Teppaitoon, P.L. Douglas, Supercritical antisolvent precipitation of andrographolide from Andrographis paniculata extracts: Effect of pressure, temperature and CO_2 flow rate. Powder Technol. **200**, 246–253 (2010)

15. ICH, International Conference on Harmonization (ICH) of Technical Requirements for the Registration of Pharmaceuticals for Human Use. Guideline for Residual Solvents Step 4 (1997)

16. R. Adami, E. Reverchon, E. Järvenpää, R. Huopalahti, Supercritical AntiSolvent micronization of nalmefene HCl on laboratory and pilot scale. Powder Technol. **182**, 105–112 (2008)

17. M.-S. Kim, S. Lee, J.-S. Park, J.-S. Woo, S.-J. Hwang, Micronization of cilostazol using supercritical antisolvent (SAS) process: effect of process parameters. Powder Technol. **177**, 64–70 (2007)

18. E. Reverchon, G. Caputo, I. De Marco, Role of phase behavior and atomization in the supercritical antisolvent precipitation. Ind. Eng. Chem. Res. **42**, 6406–6414 (2003)

19. Y. Bakhbakhi, S. Alfadul, A. Ajbar, Precipitation of Ibuprofen Sodium using compressed carbon dioxide as antisolvent, European journal of pharmaceutical sciences: official journal of the European Federation for. Pharm. Sci. **48**, 30–39 (2013)

20. E. Reverchon, Supercritical antisolvent precipitation of micro- and nano-particles. J. Supercrit. Fluids **15**, 1–21 (1999)

21. P.J. Linstrom, W. Mallard, *NIST chemistry webbook* (National Institute of Standards and Technology Gaithersburg, MD, 2001)

22. P. Pathak, M.J. Meziani, T. Desai, Y.-P. Sun, Formation and stabilization of ibuprofen nanoparticles in supercritical fluid processing. J. Supercrit. Fluids **37**, 279–286 (2006)

23. Y. Li, D.J. Yang, S.L. Chen, S.B. Chen, A.S.C. Chan, Comparative physicochemical characterization of phospholipids complex of puerarin formulated by conventional and supercritical methods. Pharm. Res. **25**, 563–577 (2008)

24. D.H. Won, M.S. Kim, S. Lee, J.S. Park, S.J. Hwang, Improved physicochemical characteristics of felodipine solid dispersion particles by supercritical anti-solvent precipitation process. Int. J. Pharm. **301**, 199–208 (2005)

25. R.E. Gordon, S.I. Amin, Crystallization of ibuprofen, Patent US 4476248 A, Google Patents (1984)

Chapter 3
Precipitation of Particles Using Combined High Turbulence Extraction Assisted by Ultrasound and Supercritical Antisolvent Fractionation

3.1 Introduction

Annatto (*Bixa Orellana L.*) seeds contain the carotenoids bixin, norbixin, and norbixinate [1]. Annatto preparations are used to impart distinctive flavor and color to foods and are a primary colorant in dairy foods such as cheese and butter [2].

The lipid-rich fraction of annatto contains a large amount of δ-tocotrienol, which can also be effective in the prevention of lipid peroxidation; probably, the tocotrienols combined with bixin act synergistically to protect the unsaturated lipids from oxidation [3]. In addition to the presence of carotenoids and tocotrienols, annatto seed extracts also contain phenolic compounds, namely hypoletin as the major compound and a caffeoyl acid derivative as the minor derivative [4].

Industrial processes of annatto pigment extraction commonly use alkaline solutions of sodium hydroxide and potassium hydroxide. In the industrial process, bixin pigment reacts with the basic solution, modifying its structure to norbixin [5]. Besides of this, traditional methods to extract colorants are time-consuming and require relatively large quantities of solvents [6].

Ultrasound-Assisted Extraction (UAE) is a recent and versatile process employed as complement methods to extract bioactive compounds with the aid of acoustic energy and solvents to extract bioactive compounds from plant matrices [7].

Recently our research group developed a novel selective extraction method to produce active solutions with from the semi-defatted seeds of annatto entitled and High Turbulence Extraction Assisted by Ultrasound (HTEAU). The HTEAU process combines the use of two types of commercial equipments and technologies. The first is Ultra-turrax® rotor-stator technology, which produces high turbulence in the plant material bed by high extracting solvent circulation flow rate (until 2000 cm^3/min) and the second is ultrasound technology, which is recognized to improve the extraction rate by the increasing the mass transfer and possible rupture of cell wall due the formation of microcavities [8].

© The Author(s), under exclusive license to Springer Nature Switzerland AG 2019
D. T. Santos et al., *Supercritical Antisolvent Precipitation Process*,
SpringerBriefs in Applied Sciences and Technology,
https://doi.org/10.1007/978-3-030-26998-2_3

Supercritical Fluid Extraction [9] and Supercritical Antisolvent Fractionation SAF [10] are clean and efficient alternatives to fractionate target compounds from plant extracts.

The versatility of supercritical fluids in green technology has led to innovative approaches for the design of micro- and nanoparticles [11]. Numerous techniques have been developed for particle formation, encapsulation, impregnation, and drying using supercritical carbon dioxide ($SC\text{-}CO_2$), demonstrating the flexibility in such technologies while offering advantages in terms of the control of particle size, size distribution and morphology that cannot be matched by conventional technologies [12].

In this context, we combined HTEAU-SAF to obtain particles from a bioactives solution (BS) produced with annatto extracts to provide a value-added product with enhanced purity in carotenoid bixin and phenolic compounds that serves for application in food and non-food industries.

3.2 Materials and Methods

3.2.1 Raw Material

The semi-defatted seeds of annatto, variety Piave were purchased from Estação dos Grãos Ltda. (São Paulo, Brazil).

3.2.2 Experimental

3.2.2.1 Supercritical Fluid Extraction

The process for semi-defatting the seeds was performed in the SFE pilot unit (Thar Technologies, SFE-2 × 5LF-2FMC, Pittsburgh, USA), using CO_2 (99% pure, White Martins, Brazil). The following conditions were 313 K, 20 MPa, CO_2 flow of 200 g/min, and solvent to feed ratio(S/F) of 11.

3.2.2.2 High Turbulence Extraction Assisted by Ultrasound

Bioactives solution (BS) composed of annatto extracts and ethanol (Synth, lot 180675, Diadema, SP) were obtained from high turbulence extraction assisted by ultrasound (HTEAU). The HTEAU equipment consisted of an Ultra-turrax® (Ultra-turrax®, DijkstraVereenigde, IKA® mag LAB®, NJ Lelystad, Holland) extraction equipment (Fig. 3.1) combined with ultrasound equipment (Unique, Indaiatuba, Brazil).

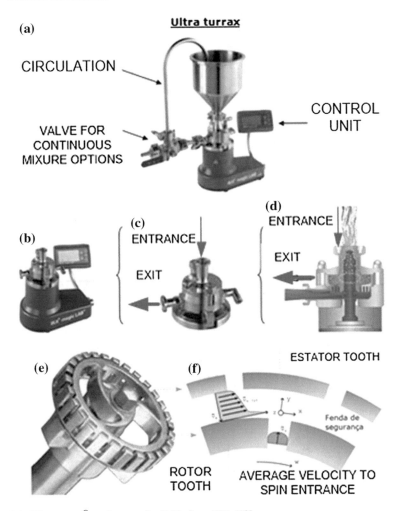

Fig. 3.1 Ultra-turrax® equipment. Available from IKA [13]

In Ultra-turrax® 2000 mL/min was used as solvent flow and rotational velocity of 26,000 min⁻¹. The equipment consisted of two continuous mixing, which is by recirculation plus the rotor speed or only by the speed of the rotor (in this study the first option was used).

In Fig. 3.1, letter (a) outlines the Ultra-turrax® equipment in the integral configuration, letter (b) represents the dispersion apparatus, letter (c) is the Micro-Plant module with its interior configuration in letter (d). The letter (e) represents the stator and rotor and in letter (f) the velocity profile is drawn in the axial and radial direction through which the SA flows. The BS was obtained with the equipment shown in Fig. 3.5. The Ultrasound was applied, by inserting the tip of this equipment into the Ultra-turrax® vessel.

Initially, the semi-defatted annatto seeds were weighed using an analytical balance (Sartorius GMBH, model A200S, ±0.0001 g, Gottingen, Germany) inside baskets of 7 cm diameter × 10 cm in length, made of polyester 150 wires per cm^2 (Silk Screen Brasil, Campinas, SP). The mass of the ethanol was also weighed, resulting in solvent to feed (S/F) ratios of 7.8, 3.9 and 1.6, respectively.

3.2.2.3   Supercritical Antisolvent Fractionation

A schematic diagram of the equipment used to perform the SAS precipitation experiments is shown in Fig. 3.2. In the CO_2 cylinder (1) exit, the SAF apparatus configuration recommends the use of a cooling bath (5) in order to control the temperature of CO_2 at the moment it is pumped by an HPLC pump (6), in order to avoid cavitation. This pump guarantees the maintenance of pressure conditions above the critical point of the SC-CO_2-organic solvent mixture meanwhile SC-CO_2 flows through a heat exchanger (7) to increase the temperature also above the critical point. In the same way, inside a recipient (8) the BS is pumped through an HPLC pump (9), which enters in contact with SC-CO_2. These fluids are mixed through an injection system (10) which is in some cases constituted by an adaption of stainless steel nozzle and connectors which facilitates the fractionation of the compounds of interest. In other cases, the injection system is defined as a three-step fractionation column, whose operational detail is described in the following section. Inside this injection system, complete solubility of the organic solvent is reached in SC-CO_2 phase (depending on the injector dimensions, a previous fractionation process is possible), following

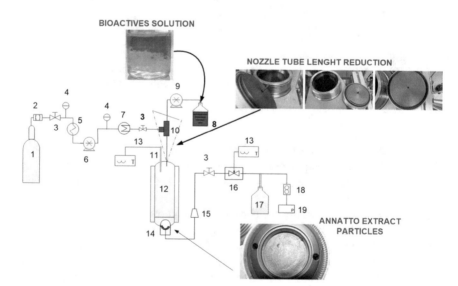

Fig. 3.2 Schematic steps for annatto extract fractionation using Supercritical Antisolvent Fractionation (SAF) process

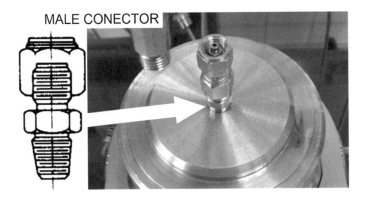

Fig. 3.3 Modification of void space in the precipitation column cover with addition of Teflon plug

its passage into the precipitation column (12), when the atomization or expansion of fluids occurs, resulting in the fractionation of the mixture through the precipitation of the purified compounds.

A stainless steel filter (14), with porosity usually ranging from 0.1 to 1.0 μm is situated in the inferior part of the precipitation column, in order to recover the highest proportion of powder solids. Leaving the column is the mixture (organic solvent + SC-CO_2 + non-precipitated compounds). A line filter (15) can help retain part of the sol-ids that were not precipitated into the precipitation column. A preliminary experiment was developed for the obtaining and concentration of precipitated particles using a SAS unit and a bioactives solution (BS) obtained from ultrasound-assisted extraction filter paper semi-defatted annatto seeds with a S/F ratio of 3.9. Afterwards, a global precipitation yield (GPY) of 20.2% was calculated from a total solid mass feed of 281.2 mg.

In preliminary assays, great quantities of solvent were used in the cleaning procedure of SAS unit because of existence of spaces inside the lids occupied by retained solids. In this case, a Teflon plug was inserted inside the lids in order to avoid undesirable solid entrances (Fig. 3.3) in difficult places for the particle collection, unnecessary solvent consumption in the lids washing, contamination of subsequent procedures with residual material from predecessor procedures, and loses of material for the calculation of GPY in each experimental procedure. This modification was performed in both lids from the precipitation vessel. The filled space volume was 0.9 mL for each lid.

3.2.3 Chemical Composition

3.2.3.1 Phenolics Compounds

The total phenolic compound on precipitated particles was determined using the Folin–Ciocalteu reagent, according to the protocol proposed by Singleton et al. [14].

3.2.3.2   Bixin

Bixin was quantified by spectrophotometry. The absorbance was measured in a spectrophotometer (FEMTO model 800XI, São Paulo, Brazil) at 487 nm according to the protocol proposed elsewhere [15].

3.2.4   Scanning Electron Microscopy

The structure of annatto particles was examined using scanning electron microscopy (SEM). The samples were applied to circular aluminum stubs with double carbon sticky tape and coated with 200 Å of gold using a sputter coater (EMITECH, K450, Kent, United Kingdom). The micrographs with magnification of 300 × were obtained using scanning electron microscope (Leo 440i, Cambridge, England) at an accelerating potential of 20 kV and current of 100 pA.

3.3   Results and Discussion

3.3.1   Global Precipitation Yield and Particle Distribution in SAF System

Experimental design consists of 8 experiments (Table 3.1). Studied variable was pressure (10 and 12 MPa), CO_2 flow (500 and 800 g/h) and bioactives solution flow (0.5 and 1 mL/min). The experiments were performed as 40 °C and nozzle length of 6.6 cm.

Table 3.1 Experimental design and precipitation yields

EXP	Pressure (MPa)	CO_2 flow (g/h)	BS flow(g/mL)	GPY (%)
8	−1	−1	−1	80.5
7	−1	1	−1	77.2
4	−1	−1	1	78.6
3	−1	1	1	78.2
6	1	−1	−1	75.6
5	1	1	−1	72.2
2	1	−1	1	74.1
1	1	1	1	74.6

Processing condition with lower global precipitation yield (GPY) corresponds to 12 MPa, 800 CO_2 g/h and 0.5 mL BS/min, i.e., experiment 5. Precipitated particles in base filter from this experiment were analyzed by scanning electron microscopy (SEM).

Highest GPY was obtained at 10 MPa, 500 g CO_2/h and 0.5 BS mL/min (Fig. 3.4), i.e., in the lowest experimental conditions (experiment 8), because of lowest flow of CO_2 and BS minimize the dragging of particles out the precipitation vessel, similarly to that obtained elsewhere for turmeric extracts coprecipitated with polyethylene glycol in compressed CO_2 [16].

In the range of pressures studied, it was observed that an increase in pressure produced a decrease of GPY in the precipitation, on the contrary to those reported elsewhere [17].

According to Table 3.2, increasing of pressure had a significant effect ($p_{value} = 0.039$) on the GPY. Lowest GPY values were obtained at 12 MPa.

Mass of particles distribution in the precipitation system using a 6.6 cm capillary nozzle is showed on Fig. 3.5.

Highest proportion of particles was observed on the base of vessel (40%), while the lowest proportions were found in the walls of vessel (10%). Distribution of particles did not have the same behavior of bixin distribution. Particle distribution in the whole SAS systems in all extraction conditions was evaluated considering the condition of 10 MPa (Fig. 3.6). It was observed that 40% of total feed bioactives were found in the base of vessel and 10% at the wall, however with higher bixin content.

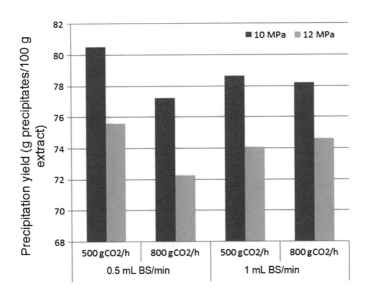

Fig. 3.4 Effect of pressure, CO_2 flow and BS flow rate on precipitation yield

Table 3.2 ANOVA for global precipitation yield (GPY)

Variable	FD	SQ	QM	F	P
Pressure	1	40.682	40.682	259.77	0.039
BS flow	1	5.302	5.302	33.85	0.108
CO_2 flow	1	0.0004	0.0004	0.00	0.967
Pressure × BSflow	1	0.097	0.097	0.62	0.575
Pressure × CO_2 flow	1	0.381	0.381	2.43	0.363
BS flow × CO_2 flow	1	5.764	5.764	36.80	0.104
Error	1	0.157	0.157		
Total	7	52.381			

*FD: Freedom degree; SQ: Square; QM: Square mean; F: F statistical; P: *p*-value

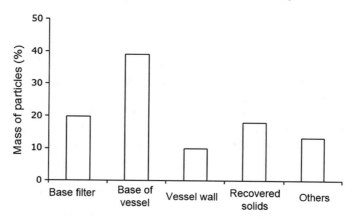

Fig. 3.5 Distribution of particle weight (g/100 g bioactive) in the SAS system using the shortest capillary nozzle length 6.6 cm

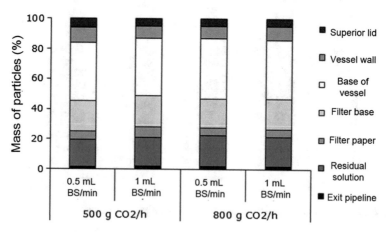

Fig. 3.6 Distribution of the mass of the particles in the SAS precipitation system at 10 MPa

3.3.2 Total Phenolics and Bixin Distribution in SAF System

Supercritical antisolvent fractionation combined with HTEAU method increased the content of total phenols from semi-defatted annatto seeds extracts in approximately 95%, when compared to HTEAU alone [8]. It was observed that highest concentration of phenolics was found in the base of the precipitation vessel, considering the condition of 10 MPa (Fig. 3.7).The content of total phenolics obtained in this work is lower than the 100 mg/g gallic acid obtained from the rosemary extracts coprecipitated particles [18], and similar to that found to the precipitates of yarrow [17] and mango leaves extracts [19] obtained using the same conditions, i.e., 313 K and 10 MPa.

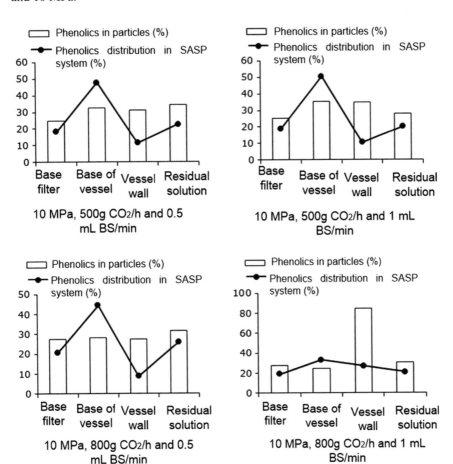

Fig. 3.7 Concentration of phenolics in the particles (mg GAE/100 mg particles) and the distribution of total phenolics precipitated during the process, by sectors of the SAS system

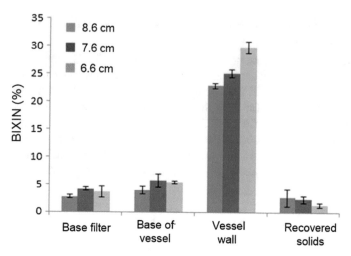

Fig. 3.8 Bixin precipitation yield (mg Bixin/100 mg of bioactives fed to the SAS unit) with three nozzle lengths

In turmeric ethanolic extracts precipitates, the use of lower pressure of 10 MPa was attributed to the highest content of curcuminoids, the phenolic compounds from the material used [20]. Considering a 6.6 cm nozzle, combined processes HTEAU-SAF increased the content of bixin in approximately 170% when compared to the HTEAU alone [8]. Bixin precipitation yield was determined according to the inside and outside parts of the precipitation vessel with the capillary nozzle lengths (Fig. 3.8). It is observed that decreasing of capillary nozzle from 8.6 to 7.6 cm, contributed to a positive effect in GPY, since the higher proportion was observed on-base filter, vessel base and vessel wall, and the lowest in the residual solution outside the vessel.

Decreasing of tube length from 7.6 to 6.6. cm enhanced the distribution of bixin particles inside the vessel because lowest concentration of bixin was determined in base filter. Furthermore, higher proportion of bixin was found at the walls of the vessel.

It is worth mentioning that this work accurately shows the distribution of the particulate content in the precipitation system. This kind of information is rarely shown in the current literature.

Effects of the interaction of CO_2 flow and vessel constituents on the content of bixin are presented in Fig. 3.9. It was observed that the reduction of CO_2 flow increases the concentration of bixin in the particles in vessel wall. Uses of highest antisolvent flows cause dragging of bioactives outside the column, because of lowest contact periods between CO_2 and the solubilized bioactives from BS. Furthermore, use of lowest CO_2 flows enhances the expansion of BS, decreasing the distance between the atomization phenom from the nozzle to the bottom of the vessel, and consequently, the atomized area has higher surface contact with the vessel wall. Contrary effect was observed to the other sectors of vessel which bixin concentration is similar considering 500 and 800 g CO_2/h.

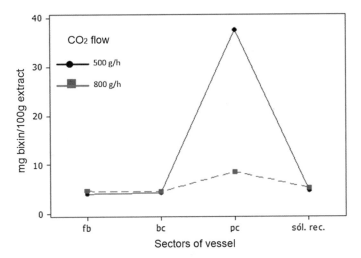

Fig. 3.9 Effects of CO_2 flow and the sectors of the vessel (fb—base filter; bc—base of vessel; pc—vessel wall; sól. rec—recovered solids) on the recovery of bixin

According to the weight distribution of bixin inside the wall of precipitation vessel (conditions of 12 MPa, 500 g CO_2/h and 1 ml BS/min) was found up to 66% of precipitated bixin probably because of its low solubility in CO_2 base, which favors its precipitation inside the vessel.

Higher CO_2 density increases the solvent power towards small molecular weight, non-polar or slightly polar compounds increases; therefore, larger quantities of undesired compounds are extracted from the starting solution and, correspondingly, the concentration of bioactives, such as phenolics, in the precipitates increases [21].

3.3.3 Scanning Electron Microscopy

According to the SEM photos, the particles present triangular morphology with sizes ranged between approximately 10 and 100 μm (Fig. 3.10), comparable with the particles from onion peels extracts [22] and differently from spherical particles from mango leaves extracts (0.02–30 μm) [23] and from olive leaves extracts 300–1060 nm [24]. In literature, the effect of pressure on particle size using carbon dioxide as antisolvent is not clear. In some cases smaller particles are formed with pressure increasing [16, 23, 25], however in other cases opposite effect is observed [26] Particles presented some aggregation (Fig. 3.10a), attributed to the prevention of diffusion between solvent and antisolvent, accelerating the aggregation of particles [27]. Microstructure of particles observed in A, B and C correspond to base filter from experiment 5, which conditions were 12 MPa, 800 g CO_2/h and 0.5 mL BS/min.

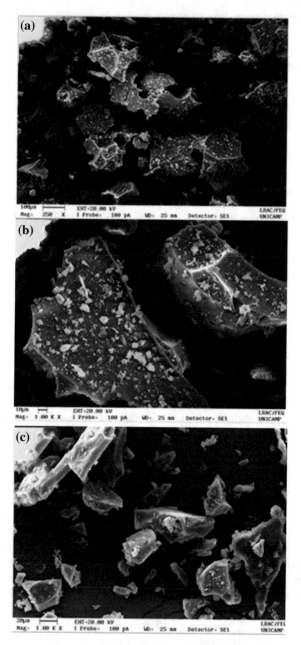

Fig. 3.10 SEM of the particles formed under the conditions of 12 MPa, 800 g CO_2/h and 0.5 mL SA/min

In literature the effect of pressure on particle size in the SAS process is not clear. In some cases smaller particles are formed as the pressure increases [28], but in other cases the opposite effect is observed [26].

3.4 Conclusions

The integrated HTEAU-SAF process resulted on annatto microparticles of triangular morphology, little aggregation state and attractive composition in terms of bixin and total phenolic content. The use of low antisolvent flow was associated with the highest precipitation yield. The configuration of nozzle was the limiting factor in the precipitation yield by this process. In this case small modifications made in the SAS precipitation unit considering filling the void space and reducing the length of the nozzle contributed to decrease the losses of product during processing, in such way that the accurate distribution of the particulate content in the precipitation system was carefully investigated. This kind of information is rarely shown in the current literature.

Acknowledgements R. Abel C. Torres thanks Capes for his doctorate assistantship. Ádina L. Santana thanks Capes (88882.305824/2013-01) for her postdoctoral financial support. M. Angela A. Meireles thanks CNPq for the productivity grant (302423/2015-0). The authors acknowledge the financial support from FAPESP (process 2015/13299-0).

References

1. M.A. Valério, M.I.L. Ramos, J.A. Braga Neto, M.L.R. Macedo, Annatto seed residue (*Bixa orellana* L.): nutritional quality. Food Sci. Technol. **35**, 326–330 (2015)
2. V. Galindo-Cuspinera, D.C. Westhoff, S.A. Rankin, Antimicrobial properties of commercial annatto extracts against selected pathogenic, lactic acid, and spoilage microorganisms. J. Food Prot. **66**(6), 1074–1078 (2003)
3. C.L.C. Albuquerque, M.A.A. Meireles, Defatting of annatto seeds using supercritical carbon dioxide as a pretreatment for the production of bixin: experimental, modeling and economic evaluation of the process. J. Supercrit. Fluids **66**, 86–95 (2012). https://doi.org/10.1016/j.supflu.2012.01.004
4. C.L.C. Albuquerque, Á.L. Santana, M.A.A. Meireles, Thin layer chromatographic analysis of annatto extracts obtained using supercritical fluid. Food Public Health **5**(4), 127–137 (2015)
5. S.C. Alcázar-Alay, J.F. Osorio-Tobón, T. Forster-Carneiro, M.A.A. Meireles, Obtaining bixin from semi-defatted annatto seeds by a mechanical method and solvent extraction: process integration and economic evaluation. Food Res. Int. **99**, 393–402 (2017). https://doi.org/10.1016/j.foodres.2017.05.032
6. M. Yolmeh, M.B. Habibi Najafi, R. Farhoosh, Optimisation of ultrasound-assisted extraction of natural pigment from annatto seeds by response surface methodology (RSM). Food Chem. **155**, 319–324 (2014). https://doi.org/10.1016/j.foodchem.2014.01.059

7. H. Bagherian, F. Zokaee Ashtiani, A. Fouladitajar, M. Mohtashamy, Comparisons between conventional, microwave- and ultrasound-assisted methods for extraction of pectin from grapefruit. Chem. Eng. Process.: Process Intensification **50**(11), 1237–1243 (2011). https://doi.org/10.1016/j.cep.2011.08.002

8. R.A.C. Torres, D.T. Santos, M.A.A. Meireles, Novel extraction method to produce active solutions from plant materials. Food Public Health **5**(2), 38–46 (2015)

9. J.C. Johner, Á.L. Santana, M.A.A. Meireles, Fractionation of annatto extracts with carbon dioxide using a home-made equipment. Food Public Health **7**(3), 69–74 (2017)

10. R.A.C. Torres, Á.L. Santana, D.T. Santos, M.A.A. Meireles, Perspectives on the application of supercritical antisolvent fractionation process for the purification of plant extracts: effects of operating parameters and patent survey. Recent Pat. Eng. **10**, 121–130 (2016)

11. J. Jung, M. Perrut, Particle design using supercritical fluids: literature and patent survey. J. Supercrit. Fluids **20**(3), 179–219 (2001). https://doi.org/10.1016/S0896-8446(01)00064-X

12. F. Temelli, Perspectives on the use of supercritical particle formation technologies for food ingredients. J. Supercrit. Fluids (2017). https://doi.org/10.1016/j.supflu.2017.11.010

13. IKA, Agitators, Batch and Inline Dispersing Machines, Laboratory Reactors and Pilot Plants (2014). http://www.ika.com.my/PDF/201007_Pilots_EN_IWK_USD_spreads_screen.pdf Accessed 23.12.2014

14. V.L. Singleton, R. Orthofer, R.M. Lamuela-Raventos, Analysis of total phenols and other oxidation substrates and antioxidants by means of Folin-Ciocalteu reagent. Oxidants and Antioxidants (1999)

15. FAO, Food and Agriculture Organization of the United Nations. In: 67th Joint FAO/WHO Expert Committee on Food Additives, vol. 3. Rome, Italy (2006), p 11

16. Á.L. Santana, M.A.A. Meireles, Coprecipitation of turmeric extracts and polyethylene glycol with compressed carbon dioxide. J. Supercrit. Fluids **125**, 31–41 (2017). https://doi.org/10.1016/j.supflu.2017.02.002

17. D. Villanueva-Bermejo, F. Zahran, D. Troconis, M. Villalva, G. Reglero, T. Fornari, Selective precipitation of phenolic compounds from *Achillea millefolium* L. extracts by supercritical anti-solvent technique. J. Supercrit. Fluids **120**, 52–58 (2017). https://doi.org/10.1016/j.supflu.2016.10.011

18. A. Visentin, S. Rodríguez-Rojo, A. Navarrete, D. Maestri, M.J. Cocero, Precipitation and encapsulation of rosemary antioxidants by supercritical antisolvent process. J. Food Eng. **109**(1), 9–15 (2012). https://doi.org/10.1016/j.jfoodeng.2011.10.015

19. M.A. Meneses, G. Caputo, M. Scognamiglio, E. Reverchon, R. Adami, Antioxidant phenolic compounds recovery from *Mangifera indica* L. by-products by supercritical antisolvent extraction. J. Food Eng. **163**, 45–53 (2015). https://doi.org/10.1016/j.jfoodeng.2015.04.025

20. J.F. Osorio-Tobón, P.I.N. Carvalho, M.A. Rostagno, A.J. Petenate, M.A.A. Meireles, Precipitation of curcuminoids from an ethanolic turmeric extract using a supercritical antisolvent process. J. Supercrit. Fluids **108**, 26–34 (2016). https://doi.org/10.1016/j.supflu.2015.09.012

21. L. Baldino, G. Della Porta, L.S. Osseo, E. Reverchon, R. Adami, Concentrated oleuropein powder from olive leaves using alcoholic extraction and supercritical CO_2 assisted extraction. J. Supercrit. Fluids **133**, 65–69 (2018). https://doi.org/10.1016/j.supflu.2017.09.026

22. G.L. Zabot, M.A.A. Meireles, On-line process for pressurized ethanol extraction of onion peels extract and particle formation using supercritical antisolvent. J. Supercrit. Fluids **110**, 230–239 (2016). https://doi.org/10.1016/j.supflu.2015.11.024

23. M.C. Guamán-Balcázar, A. Montes, C. Pereyra, E.M. de la Ossa, Precipitation of mango leaves antioxidants by supercritical antisolvent process. J. Supercrit. Fluids **128**, 218–226 (2017). https://doi.org/10.1016/j.supflu.2017.05.031

24. C. Chinnarasu, A. Montes, M.T.F. Ponce, L. Casas, C. Mantell, C. Pereyra, E.J.M. de la Ossa, Precipitation of antioxidant fine particles from *Olea europaea* leaves using supercritical antisolvent process. J. Supercrit. Fluids **97**, 125–132 (2015). https://doi.org/10.1016/j.supflu.2014.11.008

25. J.-J. Wu, L.-Y. Shen, M.-C. Yin, Y.-S. Cheng, Supercritical carbon dioxide anti-solvent micronization of lycopene extracted and chromatographic purified from *Momordica charantia* L. aril. J. Taiwan Inst. Chem. Eng. **80**, 64–70 (2017). https://doi.org/10.1016/j.jtice.2017.08.006

26. E. Reverchon, R. Adami, G. Caputo, I. De Marco, Spherical microparticles production by supercritical antisolvent precipitation: interpretation of results. J. Supercrit. Fluids **47**(1), 70–84 (2008). https://doi.org/10.1016/j.supflu.2008.06.002

27. Z. Liu, L. Yang, Antisolvent precipitation for the preparation of high polymeric procyanidin nanoparticles under ultrasonication and evaluation of their antioxidant activity in vitro. Ultrason. Sonochem. **43**, 208–218 (2018). https://doi.org/10.1016/j.ultsonch.2018.01.019

28. A. Montes, L. Wehner, C. Pereyra, E.J. Martínez de la Ossa, Precipitation of submicron particles of rutin using supercritical antisolvent process. J. Supercrit. Fluids **118**, 1–10 (2016). https://doi.org/10.1016/j.supflu.2016.07.020

Chapter 4
Recent Developments in Particle Formation with Supercritical Fluid Extraction of Emulsions Process for Encapsulation

4.1 Introduction

The microencapsulation in the food industry provides a protective barrier to sensitive target compounds, masking unpleasant tastes and smells and stabilizing and increasing the bioavailability of the bioactive compound [1]. Conventional techniques for particle formation have been proposed in the literature (spray-drying, jet milling, liquid antisolvent precipitation, solvent evaporation, emulsification, and lyophilization). However, these methods suffer from many drawbacks, mainly lack of control over particle morphology, particle size and particle size distribution (PSD), difficulty in the elimination of the solvents used and possible degradation due to high temperatures employed [2].

Aspects regarding particles are a crucial factor for processing and consumption. For instance, particles should be around 0.1–0.3 μm for intravenous delivery, 1–5 μm for inhalation delivery, and 0.1–100 μm for oral delivery [3].

Emulsion freeze-drying and solvent evaporation are expected to be fabrication techniques of drug or polymer suspensions. However, both techniques require huge amounts of organic solvent, that limits the production of suspensions, because of high costs with energy and removal of solvent, and purification of the suspension [4].

As an alternative to conventional particle formation processes, the described class of hydrophobic target compounds is suitable for crystallization by the use of through solvent extraction from oil in water (o/w) emulsion. The hydrophobic compound is first dissolved in a suitable organic solvent, and the solution is then dispersed in water, so as to form an o/w emulsion [5].

Fig. 4.1 Pressure-
temperature phase diagram
of a single substance

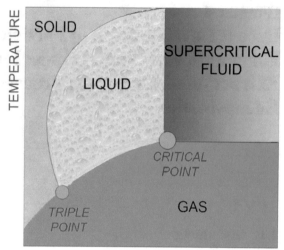

A fluid is defined as supercritical when its temperature and pressure exceed critical values (Fig. 4.1). Its solvency power is enhanced due to its higher density, which is very similar to those of liquids (0.1–0.9 g cm^{-3} at 7.5–50 MPa). The advantages on the use of CO_2 as solvents are its non-toxicity, non-flammability, low critical temperature and pressure (T_c = 304.2 K, P_c = 7.38 MPa), low cost, and it is a GRAS (Generally Regarded as Safe) solvent [6].

Supercritical fluid extraction of emulsions (SFEE) combines conventional emulsion processes with the unique properties of supercritical fluids to produce tailor-made micro- and nanoparticles. The basis of this process relies on the use of supercritical CO_2 to rapidly extract the organic solvent from an oil in water emulsion, in which a target compound and its coating polymer have been previously dissolved. Once the solvent is removed, both compounds precipitate, generating a suspension of particles in water [7].

4.2   Supercritical Fluid Extraction of Emulsions (SFEE)

In the SFEE, an oil in water (o/w) emulsion is formulated by the dissolution of target compound of interest (solute) in an organic solvent. This solution is dispersed by a surfactant material in a continuous aqueous phase. Then the emulsion is contacted with a supercritical fluid, in order to rapidly extract the organic phase from the emulsion. The supercritical fluid must be chosen to have high affinity for the organic solvent, meanwhile negligible affinity for the active compound. Due to the rapid supersaturation of the dissolution medium by the active compound, this compound is precipitated in sub-micrometric scale, encapsulated by the surfactant material [8].

The SFEE is an evolution of supercritical antisolvent (SAS) process because it is specifically suitable to encapsulate poorly water-soluble drugs in an aqueous suspension, through the combination between emulsion techniques and the SAS precipitation [9].

Emulsion techniques generally require large quantities of organic solvents, and their removal involves additional separation techniques and the use of high temperatures. In addition, SAS is not able to produce particles within the nanometric scale, and the resulting products have an increased tendency for particle agglomeration [10]. To overcome these disadvantages the removal of organic solvents during the process enables the production of nanoscale particles that improve the solubility of the aqueous solutions [11].

Perrut et al. [12] proposed and patented the processing of a water-in-oil emulsion that is the reverse of Chattopadhyay et al. [11] patent, which use $SC-CO_2$ to eliminate the organic solvent from oil-in-water emulsions. Furthermore, the process/proposed by Perrut et al. [12] uses $SC-CO_2$ to remove the organic solvent and the water.

The SFEE process developed by Ferro Corporation [13] has been validated through a myriad of successful feasibility projects and it is available for licensing. SFEE expands on established emulsion-based particle precipitation process/SCF extraction technique by combining particle engineering flexibility with the efficiency of large-scale continuous SCF extraction to produce 10 nm to 100 μm particles of small actives, lipids, polymers and some biologicals for controlled release, improved dissolution, nano-suspensions, and injectables.

Della Porta et al. [14] proposed, by using a countercurrent packed column, the SFEE process in a continuous operating mode for the production of polylactic-co-glycolic acid (PLGA) microparticles. This process design takes advantage of the large contact area between the $SC-CO_2$ and emulsion enabling the control in particle formulation into narrow size distributions in only a few minutes.

The experimental setup and principles of the SFEE process are similar as those of supercritical antisolvent (SAS), but in SFEE, the antisolvent $SC-CO_2$ remove the solvent from the droplets of an oil-in-water (O/W) or a water in oil (W/O) emulsion. The solute remains in a suspension stabilized by a surfactant agent to avoid aggregation of droplets.

The differences between the SAS and SFEE processes are as follows: (a) in SFEE, an emulsion containing the substance to be precipitated dissolved in its dispersed phase is injected, whereas in SAS, a simple solution of the substances is injected; (b) SFEE requires additional steps to produce a powdery product because an aqueous product is formed; (c) the preparation of the initial materials is more complex in SFEE; and (d) emulsion droplet size distribution is a controlling parameter in addition to the other parameters involved in the SAS process (e.g., pressure, temperature, flow rate, and solute concentration) [10].

Using the same pressure, temperature, and solution flow rate for both the SFEE and SAS methods, Shekunov et al. [15] observed a substantial difference in the resulting size and shape of the particles. The SFEE produced prismatic crystals with a volume-weighted diameter typically between 0.5 and 1 μm, whereas SAS produced longer crystal dimensions of between 20 and 200 μm and a volume-weighted diameter above 10 μm. Thus, a 10-fold reduction in the particle size was achieved using SFEE compared with the particles produced using SAS.

4.2.1 SFEE Procedures

Before initiating the SFEE process, an O/W or W/O emulsion must be prepared with the aid of surfactants. The emulsion should be stable, avoiding the coalescence phenomenon. A phase equilibria study of the complete system should be performed to know the proper operation conditions that should be carried out in the biphasic zone, to create a stable emulsion with no aggregation of particles [16].

The surfactant materials must serve in the SFEE process as surfactant to stabilize the emulsion and as coating material in the dried particles. When using a polymer devoid of emulsification properties as a coating material, such as poly-lactic-co-glycolic acid (PLGA), the use of surfactants is necessary to stabilize the emulsion. Polyvinyl alcohol (PVA) is the most popular surfactant used in the production of PLGA-stable nanoparticles in the SFEE.

There are a number of mechanisms available for the production of emulsions. High-speed stirring mixers, high-pressure homogenization, and ultrasonication have been used to form fine emulsions for use in the SFEE process [17, 18]. Microfluidization is an additional alternative for preparing submicron emulsions.

The prepared emulsion is injected in the SFEE apparatus, which can be performed in the same apparatuses used for SAS process (presented in Fig. 4.2), after slight modifications. As soon as the emulsion is introduced into the SC-CO_2 phase, the mass transfer of the organic solvent proceeds by two parallel pathways: (1) by direct extraction upon contact between SC-CO_2 and the organic phase and (2) by diffusion of the organic solvent into water followed by consequent extraction of the solvent from the aqueous phase into SC-CO_2. There is also an inverse flux of CO_2 into the droplets leading to expansion of the organic phase and creating local supersaturation and precipitation of solutes [15]. The final product of SFEE consists of aqueous micro- or nanosuspensions.

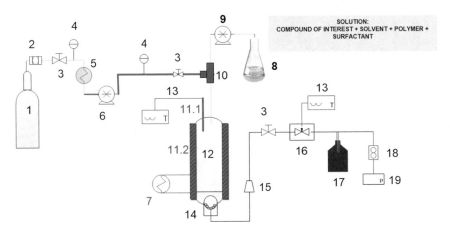

Fig. 4.2 Schematic diagram of the SFEE apparatus. 1—CO_2 cylinder; 2—CO_2 Filter; 3—Blocking valves; 4—Manometers; 5—Cooling bath; 6—CO_2 pump; 7—Heating bath; 8—solution (solute/solvent) reservoir; 9—HPLC Pump; 10—Thermocouple; 11—Precipitation vessel; 12—Temperature controllers; 13—Filter; 14—Line filter; 15—Micrometering valve with a heating system; 16—Glass flask; 17—Glass float rotameter; 18—Flow totalizer

Water can subsequently be removed by conventional drying processes, such as spray drying, lyophilization, and microwaving. The high temperature used in most conventional dryers is unsuitable for drying suspensions of some target compounds because it degrades such compounds. This step can also promote destabilization of the nanoparticles dissolved in water, increasing the particle size. The final particle size is controlled mainly by the properties of the emulsion, and not by the operating parameters of the SFEE process, such as pressure, temperature, processing time and solvent/antisolvent flow rates.

4.2.2 Applications

Supercritical Fluid Extraction of Emulsions (SFEE) is an encapsulation technology that combines conventional emulsion processes with the unique properties of supercritical fluids to produce tailored micro- and nanoparticles [7, 9]. Process optimization has been investigated for the effective encapsulation of valuable constituents, like fish oil [19, 20], pharmaceuticals [21, 22] and edible oil [23] (Table 4.1).

Table 4.1 Recent applications of SFEE

Substance	Surfactant/ Polymer	Solvent/ Antisolvent	Operational parameters	Results	References
Fish oil	Surfactant: HPMC Polymer: Polyethylene glycol 6000 (PEG 6000)	Solvent non-indentified /SC-CO_2	Pressure: 14–16 MPa Temperature: 313–333 K Actives concentration: Solution flow rate: 0.06–0.24 L h^{-1} CO_2 flow rate: non-identified	Particle size: 32.98–90.93 μm Morphology: spherical with no internal void Structure: Polymorphic nature: Residual solvent: Precipitation yield: Encapsulation efficiency: 69.55–81.75%	[19]
Fish oil	Surfactant: Tween 80 Polymer: Polycaprolactone	Acetone/SC-CO_2	Pressure: 8 MPa Temperature: 313 K Surfactant concentration: Actives concentration: Solution flow rate CO_2 flow rate: 0.6 g min^{-1}	Particle size: 6–73nm Morphology: spherical Structure: Polymorphic nature: Residual solvent: 30–19,300 ppm Precipitation yield: Precipitation efficiency: 10–40%	[20]

(continued)

Table 4.1 (continued)

Substance	Surfactant/ Polymer	Solvent/ Antisolvent	Operational parameters	Results	References
Ibuprofen	Surfactant: Chitosan Polymer: polyvinyl alcohol	Ethyl acetate-Water/SC-CO_2	Pressure: 10 MPa Temperature: 310 K Solution flow rate: 0.02 mL min^{-1} CO_2 flow rate: 30 mL min^{-1}	Particle size: 45–1591 nm Morphology: spherical Structure: Polymorphic nature: Residual solvent: Precipitation yield: Encapsulation efficiency:	[21]
Medroxyprogesterone	Surfactant: Polymer: poly(3-hydroxybutirate-co-3-hydroxyvalerate)	Dichloromethane-Poly(vinyl alcohol) PVA-Water/SC-CO_2	Pressure: 10 MPa Temperature: 313 K Polymer concentration: 8 mg mL^{-1} Actives concentration: 2% (w/w) Solution flow rate: 0.5 mL min^{-1} CO_2 flow rate: 6.3 g min^{-1}	Particle size: 850–183 nm Morphology: spherical Structure: Polymorphic nature: Residual solvent: Precipitation yield: Encapsulation efficiency: 70.3%	[22]

(continued)

Table 4.1 (continued)

Substance	Surfactant/ Polymer	Solvent/ Antisolvent	Operational parameters	Results	References
Carotenes and vitamin E from red palm oil	Surfactant: Soy lecithin, sodium caseinate, maltodextrin	Water/SC-CO$_2$	Pressure: 10–15 MPa Temperature: 313–333K Surfactant concentration: Actives concentration: Solution flow rate: 0.15 L h^{-1} CO$_2$ flow rate: 150 L h^{-1}	Particle size: 4.50–16.6 µm Morphology: spherical Structure: amorphous Polymorphic nature: Residual solvent: Precipitation yield: Precipitation efficiency: 46.3–84.6% (carotenes) and 59–95.3% (vitamin E)	[23]
Capsaicinoids from pepper oleoresin	Polymer: modified starch Hi-cap 100	Ethyl acetate/SC CO$_2$	Pressure: 9–11 MPa Temperature: 313 K Polymer concentration: 6–12 g L^{-1} Actives concentration: Hi-Cap 100/Oleoresin ratio: 0.6–1.35 Solution flow rate: 0.03–0.06 L h^{-1} CO$_2$ flow rate: 1.35 L h^{-1}	Particle size: 126–275 nm Morphology: spherical Structure: amorphous Polymorphic nature: Residual solvent: 5900–24,100 ppm Precipitation yield: Emulsification efficiency: 62–100%	[24]

(continued)

Table 4.1 (continued)

Substance	Surfactant/ Polymer	Solvent/ Antisolvent	Operational parameters	Results	References
Quercetin	Surfactant: Soy-bean lecithin Polymer: Pluronic L64®	Ethyl acetate/SC-CO_2	Pressure: 8–10 MPa Temperature: 307–313 K Polymer/Actives concentration: Solution flow rate: 0.18–0.42 L h^{-1} CO_2 flow rate: 8–12 kg h^{-1}	Particle size: 0.94–2.3 μm Morphology: semi-spherical Structure: amorphous Polymorphic nature: Residual solvent: 190–1985 mg L^{-1} Precipitation yield: Encapsulation efficiency: 78.9–98.5%	[25]
Superparamagnetic nanoparticles	Surfactant: Tween 80 Carboxybetaine-functionalized chitosan Polymer: poly-lactic-co-glycolic acid (PLGA) or polylactic acid	Ethyl acetate–Water/SC-CO_2	Pressure: 9 MPa Temperature: 311 K Liquid/gas ratio: 0.1 CO_2 flow rate: 1.4 kg h^{-1}	Particle size: 828–1114 nm Morphology: spherical Structure: Polymorphic nature: Residual solvent: Precipitation yield: Encapsulation efficiency: 1.2–5%	[26]

(continued)

Table 4.1 (continued)

Substance	Surfactant/ Polymer	Solvent/ Antisolvent	Operational parameters	Results	References
Vitamin E	Surfactant: Tween 80 Polymer: Polycaprolactone	Acetone-Water/SC-CO_2	Pressure: 8 MPa Temperature: 313 K Polymer concentration: Actives concentration: Solution flow rate: 3.9 g min^{-1} CO_2 flow rate: 10–30 g min	Particle size: 9–84 nm Morphology: spherical Structure: Polymorphic nature: Residual solvent: 1400 ppm Precipitation yield: Encapsulation efficiency: > 70%	[27]

4.2.3 Effects of Operational Conditions in SFEE Process

4.2.3.1 Temperature and Pressure

Temperature may change the hydrophilic character of the surfactant, or even the loss of its surfactive character [28]. The stability of the emulsion may reduce when the pressure is increased. Although the temperature has a minor effect, stability is related to the creaming effect.

The operating pressure and temperature conditions are selected to facilitate the maximum extraction of the organic solvent of the emulsion with minimal loss of the solute and polymer due to dissolution in CO_2 and to avoid the loss of any emulsion that may wash out in the CO_2 stream. For instance, high temperatures and pressures modify the surfactant-organic phase interactions, affecting the stability of the emulsion [11, 14].

Depending on the system studied, process conditions should be applied carefully. For instance, Falco et al. [29], Della Porta et al. [30] and Cricchio et al. [26] performed experiments with poly-lactic-co-glycolic acid (PLGA) emulsions at 80 bar and 310 K to enhance the extraction of the oily dispersed phase of the emulsion. These conditions assured the complete miscibility of ethyl acetate in SC-CO_2 whereas, the continuous phase of the emulsion (i.e., EA-saturated aqueous phase) is slightly soluble in SC-CO_2. Moreover, using this process conditions, the difference in density between the emulsion and SC-CO_2 is very large (~1 g/cm^3 for the liquid phase, 0.310 g/cm^3 for CO_2), favoring the counter-current operation in the packed column.

4.2.3.2 Emulsion Properties

The primary parameters responsible for particle size control are the emulsion droplet size, solute/solution concentration and organic solvent content in the emulsion [15]. The stability of the emulsion is associated with interfacial tension. For instance, increasing the interfacial tension increases the mass transfer of CO_2 to the drop, and the emulsion becomes destabilized.

Contact between the emulsion and CO_2 to achieve precipitation through the anti-solvent effect must occur over a short period of time to minimize the emulsion destabilization prior to precipitation. However, the removal of the remaining organic solvent may be slower because emulsion destabilization is no longer an issue after the particles have been produced [31].

The increase in solvent concentration in the emulsion increase aggregation of droplets, resulting in larger particles. The increase in particle size based on the solute concentration is likely due to an increase in the surface tension of the organic solution, resulting also in emulsions with larger droplets.

The increasing of surfactant concentration decreases the particle size. However, continuously increasing the amount of surfactant in water decreases the polydispersity index of the final product [17].

The average particle size may also decrease with an increased emulsion stirring rate, whereas the particle size distributions generally became narrower [14].

4.2.3.3 CO_2 and Emulsion Flow Rate

The CO_2 flow rate during SFEE process is directly related to the rate of solvent extraction from the emulsion droplet and solute/polymer losses, which have a significant effect on the encapsulation efficiency and final particle size [18].

Higher CO_2 flow rate induces the emulsion wash out from the extraction vessel and part of the water as well as some particles might be lost in the downstream separator (Della Porta et al., 2008). High emulsion flow rate induces high encapsulation efficiency when processing a solute with low solubility in SC-CO_2. However, when the solute has high solubility in CO_2 the encapsulation efficiency is decreased, due to dissolution in the CO_2 plus solvent mixture.

In the nanoencapsulation of vitamin E in polycaprolactone was observed that an increase in CO_2 flow rate led to a higher solvent (acetone) extraction rate. Larger flow rates enhanced Reynolds numbers and superficial solvent velocity, which benefited turbulence and external mass transfer. On the other hand, larger flow rates reduced contact time for acetone extraction [28].

4.3 Concluding Remarks

Supercritical Fluid Extraction of Emulsion (SFEE) was recently proposed for the production of biopolymer particles by several authors from *oil-in-water* emulsions. From a scientific point of view, particle design using the SAS precipitation and SFEE process are sustainable options to obtaining particles with no toxicity, besides controlled particle size and morphology, narrow size distribution and acceptable residual organic solvent content.

The main advantages of these processes are: (a) the processes can take place at near ambient temperatures, thus avoiding thermal degradation of the processed solutes; (b) they are adaptable for continuous operations being possible large-scale mass production of fine particles; (c) they allow solvent (CO_2 and organic solvent) recycling. Few reports have compared the particles obtained by both processes, otherwise, it is expected that smaller particle size is obtained by SFEE process with the adequate selection of the process for water removal. On the other hand, several studies have demonstrated the same trend: solute processing by SAS or SFEE improves its dissolution rate.

The most obvious drawback of SFEE is that the resulting suspension is an aqueous product instead of dry particles. Additional steps are required to produce a powdery product if required, which can lead to an increase in particle sizes due to agglomeration. Another limitation of this technique is that it is only suitable for the encapsulation of hydrophobic compounds. Differently of SAS, SFEE is not a one-step process.

A previous step is necessary for obtaining an emulsion and a step after SFEE have to be added to produce a dry product if such product specification is required. An advantage for SFEE implementation in the industry site is that both steps could be done in the already available emulsification and drying equipment, sharing part of the possible existing infrastructure.

References

1. V. Nedovic, A. Kalusevic, V. Manojlovic, S. Levic, B. Bugarski, An overview of encapsulation technologies for food applications. Proc. Food Sci. **1**, 1806–1815 (2011). https://doi.org/10.1016/j.profoo.2011.09.265
2. W. Wang, G. Liu, J. Wu, Y. Jiang, Co-precipitation of 10-hydroxycamptothecin and poly (l-lactic acid) by supercritical CO_2 anti-solvent process using dichloromethane/ethanol co-solvent. J. Supercrit. Fluids **74**, 137–144 (2013). https://doi.org/10.1016/j.supflu.2012.11.022
3. P. York, U.B. Kompella, B.Y. Shekunov, Supercritical Fluid Technology for Drug Product Development (CRC Press, 2004)
4. Y. Murakami, Y. Shimoyama, Supercritical extraction of emulsion in microfluidic slug-flow for production of nanoparticle suspension in aqueous solution. J. Supercrit. Fluids **118**, 178–184 (2016). https://doi.org/10.1016/j.supflu.2016.08.009
5. J. Kluge, L. Joss, S. Viereck, M. Mazzotti, Emulsion crystallization of phenanthrene by super-critical fluid extraction of emulsions. Chem. Eng. Sci. **77**, 249–258 (2012). https://doi.org/10.1016/j.ces.2011.12.008
6. G. Brunner, Supercritical fluids: technology and application to food processing. J. Food Eng. **67** (2005). https://doi.org/10.1016/j.jfoodeng.2004.05.060
7. C. Prieto, C.M.M. Duarte, L. Calvo, Performance comparison of different supercritical fluid extraction equipments for the production of vitamin E in polycaprolactone nanocapsules by supercritical fluid extraction of emulsionsc. J. Supercrit. Fluids **122**, 70–78 (2017). https://doi.org/10.1016/j.supflu.2016.11.015
8. G. Lévai, J.Q. Albarelli, D.T. Santos, M.A.A. Meireles, Á. Martín, S. Rodríguez-Rojo, M.J. Cocero, Quercetin loaded particles production by means of supercritical fluid extraction of emulsions: process scale-upstudy and thermo-economic evaluation. Food Bioprod. Process. **103**, 27–38 (2017). https://doi.org/10.1016/j.fbp.2017.02.008
9. G. Lévai, Á. Martín, S.R. Rojo, M.J. Cocero, T.M. Fieback, Measurement and modelling of mass transport properties during the supercritical fluid extraction of emulsions. J. Supercrit. Fluids **129**, 36–47 (2017). https://doi.org/10.1016/j.supflu.2017.01.015
10. M.J. Cocero, Á. Martín, F. Mattea, S. Varona, Encapsulation and co-precipitation processes with supercritical fluids: fundamentals and applications. J. Supercrit. Fluids **47**(3), 546–555 (2009). https://doi.org/10.1016/j.supflu.2008.08.015
11. P. Chattopadhyay, B.Y. Shekunov, J.S. Seitzinger, R. Huff, Particles from Supercritical Fluid Extraction of Emulsion. USA Patent (2004)
12. M. Perrut, J. Jung, F. Leboeuf, Method for Obtaining Solid Particles from at Least a Water Soluble Product. USA Patent (2004)
13. Ferro, Ferro Corporation (2019). http://www.ferro.com
14. G.D. Porta, N. Falco, E. Reverchon, Continuous supercritical emulsions extraction: a new technology for biopolymer microparticles production. Biotechnol. Bioeng. **108**(3), 676–686 (2011). https://doi.org/10.1002/bit.22972
15. B.Y. Shekunov, P. Chattopadhyay, J. Seitzinger, R. Huff, Nanoparticles of poorly water-soluble drugs prepared by supercritical fluid extraction of emulsions. Pharm. Res. **23**(1), 196–204 (2006)

16. A. Tabernero, E.M.M. del Valle, M.A. Galán, Supercritical fluids for pharmaceutical particle engineering: methods, basic fundamentals and modelling. Chem. Eng. Process. **60**, 9–25 (2012)
17. M. Furlan, J. Kluge, M. Mazzotti, M. Lattuada, Preparation of biocompatible magnetite–PLGA composite nanoparticles using supercritical fluid extraction of emulsions. J. Supercrit. Fluids **54**(3), 348–356 (2010)
18. D.T. Santos, Á. Martín, M.A.A. Meireles, M.J. Cocero, Production of stabilized submicrometric particles of carotenoids using supercritical fluid extraction of emulsions. J. Supercrit. Fluids **61**, 167–174 (2012)
19. F.T. Karim, K. Ghafoor, S. Ferdosh, F. Al-Juhaimi, E. Ali, K.B. Yunus, M.H. Hamed, A. Islam, M. Asif, M.Z.I. Sarker, Microencapsulation of fish oil using supercritical antisolvent process. J. Food Drug Anal. **25**(3), 654–666 (2017). https://doi.org/10.1016/j.jfda.2016.11.017
20. C. Prieto, L. Calvo, The encapsulation of low viscosity omega-3 rich fish oil in polycaprolactone by supercritical fluid extraction of emulsions. J. Supercrit. Fluids **128**, 227–234 (2017). https://doi.org/10.1016/j.supflu.2017.06.003
21. Y. Murakami, Y. Shimoyama, Production of nanosuspension functionalized by chitosan using supercritical fluid extraction of emulsion. J. Supercrit. Fluids **128**, 121–127 (2017). https://doi.org/10.1016/j.supflu.2017.05.014
22. W.M. Giufrida, V.F. Cabral, L. Cardoso-Filho, Conti D. dos Santos, V.E.B. de Campos, S.R.P. da Rocha, Medroxyprogesterone-encapsulated poly(3-hydroxybutirate-co-3-hydroxyvalerate) nanoparticles using supercritical fluid extraction of emulsions. J. Supercrit. Fluids **118**, 79–88 (2016). https://doi.org/10.1016/j.supflu.2016.07.026
23. W.J. Lee, C.P. Tan, R. Sulaiman, R.L. Smith, G.H. Chong, Microencapsulation of red palm oil as an oil-in-water emulsion with supercritical carbon dioxide solution-enhanced dispersion. J. Food Eng. **222**, 100–109 (2018). https://doi.org/10.1016/j.jfoodeng.2017.11.011
24. ACd Aguiar, L.P.S. Silva, CAd Rezende, G.F. Barbero, J. Martínez, Encapsulation of pepper oleoresin by supercritical fluid extraction of emulsions. J. Supercrit. Fluids **112**, 37–43 (2016). https://doi.org/10.1016/j.supflu.2016.02.009
25. G. Lévai, Á. Martín, A. Moro, A.A. Matias, V.S.S. Gonçalves, M.R. Bronze, C.M.M. Duarte, S. Rodríguez-Rojo, M.J. Cocero, Production of encapsulated quercetin particles using supercritical fluid technologies. Powder Technol. **317**, 142–153 (2017). https://doi.org/10.1016/j.powtec.2017.04.041
26. V. Cricchio, M. Best, E. Reverchon, N. Maffulli, G. Phillips, M. Santin, G. Della Porta, Novel superparamagnetic microdevices based on magnetized PLGA/PLA microparticles obtained by supercritical fluid emulsion and coating by carboxybetaine-functionalized chitosan allowing the tuneable release of therapeutics. J. Pharm. Sci. **106**(8), 2097–2105 (2017). https://doi.org/10.1016/j.xphs.2017.05.005
27. C. Prieto, L. Calvo, C.M.M. Duarte, Continuous supercritical fluid extraction of emulsions to produce nanocapsules of vitamin E in polycaprolactone. J. Supercrit. Fluids **124**, 72–79 (2017). https://doi.org/10.1016/j.supflu.2017.01.014
28. C. Prieto, L. Calvo, Supercritical fluid extraction of emulsions to nanoencapsulate vitamin E in polycaprolactone. J. Supercrit. Fluids **119**, 274–282 (2017). https://doi.org/10.1016/j.supflu.2016.10.004
29. N. Falco, E. Reverchon, G. Della Porta, Injectable PLGA/hydrocortisone formulation produced by continuous supercritical emulsion extraction. Int. J. Pharm. **441**(1), 589–597 (2013). https://doi.org/10.1016/j.ijpharm.2012.10.039
30. G. Della Porta, F. Castaldo, M. Scognamiglio, L. Paciello, P. Parascandola, E. Reverchon, Bacteria microencapsulation in PLGA microdevices by supercritical emulsion extraction. J. Supercrit. Fluids **63**, 1–7 (2012). https://doi.org/10.1016/j.supflu.2011.12.020
31. F. Mattea, Á. Martín, A. Matías-Gago, M.J. Cocero, Supercritical antisolvent precipitation from an emulsion: β-carotene nanoparticle formation. J. Supercrit. Fluids **51**(2), 238–247 (2009)

Chapter 5
Supercritical Fluid Extraction of Emulsion Obtained by Ultrasound Emulsification Assisted by Nitrogen Hydrostatic Pressure Using Novel Biosurfactant

5.1 Introduction

Recently, there has been increasing interest within the industry in replacing synthetic ingredients with natural "label-friendly" alternatives [1]. The choice of the surfactant and method used to produce emulsions is crucial. The formation of an emulsion includes the mixing of the immiscible liquids and the time for surfactant molecules to organize at the interface of the two phases [2]. Saponins are natural surface-active substances (surfactants) present in more than 500 plant species. Due to the presence of a lipid-soluble aglycone and water-soluble sugar chain(s) in their amphiphilic structure, saponins are surface-active compounds with detergent, wetting, emulsifying and foaming properties [3].

Ultrasound is one means among others of mechanically producing emulsions [2]. On the other hand, few studies have been evaluated the ultrasound-based emulsification process under pressure. In this work, the influence of hydrostatic pressure levels (up to 10 bars applying nitrogen), oily phase type and surfactant type were evaluated. In addition, the effect of saponin-rich extract solution concentration obtained from Brazilian Ginseng (*Pfaffia glomerata*) roots using hot pressurized water as extracting solvent was also evaluated to further processing of this emulsion by Supercritical Fluid Extraction of Emulsions (SFEE) process, using an oily bixin-rich extract from annatto seeds (*Bixa orellana* L.) as core material (extracting solution from hot ethyl acetate pressurized liquid extraction).

5.2   Materials and Methods

5.2.1   *Materials*

Clove essential oil used in this study was obtained using supercritical carbon dioxide as extracting solvent using a pilot-scale equipment. Operational conditions selected were 40 °C/15 MPa, The composition of this essential oil is presented elsewhere [4]. The n-octenyl succinic anhydride (OSA)-modified starch (HICAP), provided by National Starch Food Innovation (Hamburg, Germany) was used as surfactant material.

The evaluated alternative surfactant nonpurified aqueous extract from *Pfaffia glomerata* roots were obtained employing pressurized water as extracting solvent as described following. Dried and milled pieces of Brazilian ginseng roots (4.5 g) were placed in a 6.57-cm^3 extraction cell (Thar Designs, Pittsburg, USA) containing a sintered metal filter at the bottom and upper parts. The diagram of the PLE system is shown in Fig. 5.1. The cell containing the sample was heated, filled with extraction solvent (distilled water) and then pressurized. The sample was placed in the heating system for 6 min to ensure that the extraction cell would be at the desired temperature (60 °C) during the filling and pressurization procedure. After pressurization, the sample with pressurized solvent was kept statically at the desired pressure (120 atm) for the desired time (4 min). Thereafter, the backpressure regulator (BPR)

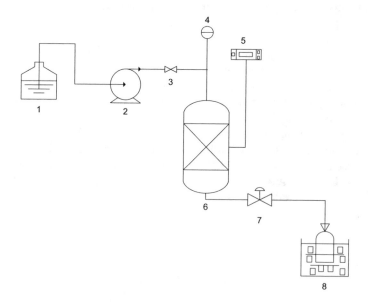

Fig. 5.1 Schematic diagram of the PLE apparatus. 1 Solvent reservoir; 2 HPLC Pump; 3 Blocking valve; 4 Manometer; 5 Temperature controller; 6 Extractor column; 7 Back pressure regulator; 8 Sampling bottle

valve (Tescom, 26-1761-24-161, ELK River, USA) was carefully opened, keeping the pressure at an appropriate level for the desired flow (1.4 cm^3/min), to rinse the extraction cell with fresh extracting solvent for 14 min (dynamic extraction time). After pressurized liquid extraction (PLE), the extracts were rapidly cooled to −5 °C in ice water using glass flasks to prevent extract degradation. The aqueous extract was stored in freezer (Metalfrio, model DA 420, São Paulo, Brazil) at −10 °C until the evaluation of its direct use as alternative surfactant. To estimate the concentration of solids in the aqueous extract it was freeze-dried for 5 days at 60–100 μHg and at −50 °C (Liobras, Liotop L101, São Carlos, Brazil).

5.2.2 Oily Bixin-Rich Extract Production

Four and one-half grams of annatto seeds were placed in the same PLE system described before (Fig. 5.1), following similar extraction procedure. The cell containing the sample was heated at 80 °C by an electrical heating jacket for 6 min to ensure that the sample reaches thermal equilibrium, and then filled with Ethyl acetate PA ACS (Merck KGaA, K40235423, Darmstadt, Germany) and pressurized at 120 atm. A volume of 18 mL of extract was collected into an amber glass vial immersed in ice bath at ambient pressure to prevent bixin degradation. The oily extract was stored in freezer (Metalfrio, model DA 420, São Paulo, Brazil) at −10 °C until the evaluation of its direct use as core material in the emulsification system proposed in this study (using nonpurified aqueous extract from *Pfaffia glomerata* roots as biosurfactant). To estimate the concentration of solids in the oily extract the solvent (ethyl acetate) was removed by using a vacuum-equipped rota-evaporator (Laborota 4001 WB, Heidolph e CH-9230, Buchi, Flawil, Switzerland) with water bath set at 40 °C.

5.2.3 Emulsion Preparation and Characterization

Oil-in-water emulsions were prepared by ultrasonication under hydrostatic pressure by applying nitrogen pressure in a high-pressure home-made system, which allowed the inclusion of a 125 ml becker containing the materials to be emulsified always in the same set-up. Thus, the emulsification was done by the contact of an ultrasound probe in the mixture prepared as described below under a nitrogen atmosphere regulated its pressure by micrometering valves.

A surfactant suspension was initially prepared by dispersing the modified starch in distilled water (100 g/ml). This first step was not necessary when using Brazilian ginseng roots aqueous extracts as surfactant. Afterwards clove essential oil, limonene, soy oil or oily bixin-rich extract in the specified ratio was gradually added to the suspension (volume fraction of oily phase of 2.5%). This resulting mixture (50 ml) was then subjected to ultrasonication using a high-grade titanium alloy probe (UNIQUE,

Indaiatuba, Brazil) for ultrasound generation at a power input of 800 W and a frequency of 20 kHz during 2 min. Proper mixing of the phases gives good emulsion which is white in color. The emulsion droplet size (expressed as the Sauter, [D 3, 2]) was determined by the laser light scattering method using Mastersizer 2000 with a Hydro 2000MU as dispersion unit.

5.2.4 Supercritical Fluid Extraction of Emulsions (SFEE) Process Description

The emulsion obtained by ultrasonication under hydrostatic pressure process containing oily bixin-rich extract obtained using PLE technique and ethyl acetate as extracting solvent and Brazilian ginseng roots aqueous extracts as surfactant was added to a SFFE equipment in order to eliminate the organic solvent. A schematic diagram of the SFEE apparatus used is shown in Fig. 5.2.

Carbon dioxide (99% CO_2, Gama Gases Especiais Ltd., Campinas, Brazil) was employed as antisolvent due to its very low solubility as a compressed fluid in the temperature and pressure ranges investigated. The CO_2 was cooled at -10 °C by a thermostatic bath (Marconi, MA-184, Piracicaba, Brazil) to ensure the liquefaction of the gas and to prevent cavitation, then it was pumped by an air-driven liquid pump (Maximator Gmbh, PP 111, Zorge, Germany) to the precipitation vessel (volume

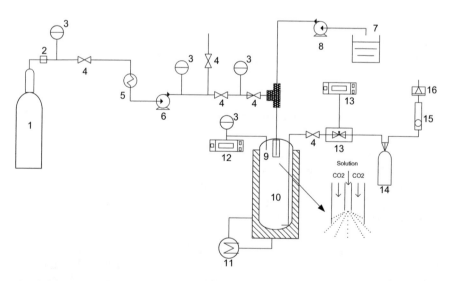

Fig. 5.2 Schematic diagram of the SFEE apparatus. 1 CO_2 Cylinder; 2 CO_2 Filter; 3 Manometers; 4 Blocking Valves; 5 Thermostatic bath; 6 CO_2 Pump; 7 Emulsion reservoir; 8 HPLC pump; 9 Thermocouple; 10 Precipitation vessel; 11 Heating bath; 12 Temperature controllers; 13 Micrometric valve with a heating system; 14 Glass flask; 15 Glass float rotameter; 16 Flow totalizer

of 500 mL; 6.8 cm internal diameter) via a nozzle. The nozzle consists of a 1/16-in. tube (inner diameter [i.d.]: 177.8 mm) for the solution extract/solvent, placed inside a 1/8-in. tube for the CO_2. Once the precipitation vessel reach temperature, pressure and CO_2 flow rate of 40 °C, 100 atm and 0.3 mL min^{-1}, respectively, the emulsion prepared by the proposed process was introduced into the vessel by a high-performance liquid chromatography (HPLC) pump (Thermoseparation Products, ConstaMetric 3200 P/F, Fremont, USA) through the coaxial annular passage of the atomizer. The vessel temperature was maintained constant at 40 °C by a heating water bath (Marconi, MA 127BO, Piracicaba, Brazil). CO_2 flow rate was measured using a glass float rotameter (ABB, 16/286A/2, Warminster, USA) coupled to a flow totalizer (LAO, G0, 6, Osasco, Brazil). When 20 mL of emulsion has been injected, the HPLC pump was stopped and only the flow of CO_2 was maintained for more 10 min for the complete removal of the solvent from the precipitator, which was proven necessary by preliminary experiments. The produced suspension was maintained at the bottom of the vessel while the fluid mixture (CO_2 plus ethyl acetate) exited the vessel and flowed to a second vessel (100-mL glass flask) connected after the micrometric valve. A heating system maintained at 120 °C was used to heat the micrometric valve to avoid the Joule–Thompson freezing effect that can lead to clogging of the throttling device during SFEE procedure. In the end, the high-pressure vessel was slowly depressurized to atmospheric pressure and suspension without the ethyl acetate solvent was collected and stored in the dark in a domestic freezer (Double Action, Metalfrio, São Paulo, Brazil) at −10 °C until subsequent analysis and characterization.

5.2.5   SFEE Product Characterization

Gas chromatography (GC) was used to determine the residual amount of ethyl acetate in the emulsion produced. The residual solvent was analyzed using a Shimadzu gas chromatograph (GC-17-A, Kyoto, Japan) equipped with flame ionization detection (FID) system. The sample was dissolved in 1 mL of toluene with the aid of an ultrasonic bath (Unique, Max Clean 1400, 40 Hz, Indaiatuba, Brazil). Sample solutions (1 μL) were introduced by direct injection on a Zebron ZB-5 capillary column from Phenomenex (30 m × 0.25 mm and 0.25 μm). The other conditions were: injection temperature of 220 °C; detector temperature of 240 °C; helium flow rate of 28 mL min^{-1} and split ratio of 1:20. Helium served as the carrier gas, and the analysis was performed using an oven temperature of 40 °C with a ramp of 20 °C min^{-1} until a temperature of 180 °C was reached. The data were quantified using a calibration curve that was constructed by measuring different known concentrations of ethyl acetate in toluene.

5.3 Results and Discussion

Using clove essential oil obtained employing supercritical carbon dioxide as extracting fluid as model oily phase, oil-in-water emulsions (droplet size in the range of 191–896 nm) were formed using N-octenyl succinic anhydride (OSA)-modified starch as surfactant material. Low overpressures (ΔP) of up to 0.75 bar improved the emulsification process, nevertheless, a further increase of pressure had a negative influence. Furthermore, when the pressure was elevated up to 2 bars, formation of emulsions stopped corroborating other authors findings (Fig. 5.3) [5].

Also using N-octenyl succinic anhydride (OSA)-modified starch (HICAP) as surfactant material, Figs. 5.4 and 5.5 show, respectively, that the same behaviour observed when Supercritical clove oil was used as oily phase was observed when Limonene and Soybean oil were tested, indicating that independently of the oily phase the optimum pressure added to the emulsion system should be found with only a few increments (0.5–0.75 atm).

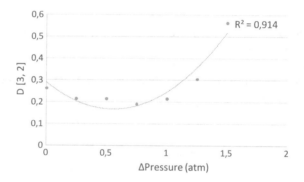

Fig. 5.3 Influence of overpressure (ΔP) during ultrasound emulsification on mean droplet diameter using Supercritical clove oil as oily phase, volume fraction of 2.5% (HICAP as surfactant—100 g/l)

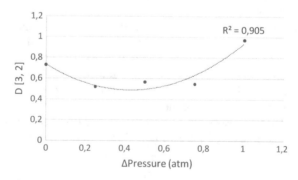

Fig. 5.4 Influence of overpressure (ΔP) during ultrasound emulsification on mean droplet diameter using Limonene as oily phase, volume fraction of 2.5% (HICAP as surfactant—100 g/l)

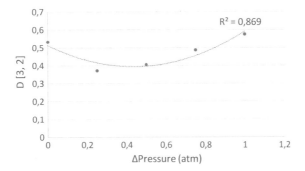

Fig. 5.5 Influence of overpressure (ΔP) during ultrasound emulsification on mean droplet diameter using Soybean oil as oily phase (HICAP as surfactant—100 g/l)

Aiming at evaluating the potentiality of the use of aqueous extract from Brazilian ginseng roots as an alternative biosurfactant in this alternative emulsification process we tested the effect surfactant concentration (Fig. 5.6) for optimization purposes and observed that a similar parabolic trend was also observed when this alternative biosurfact concentration was increased. In addition, it could be observed that the adjusted r-squared of this plotted trend is lower than the others, indicating maybe this occurred due to the use of a nonpurified aqueous extract. This extract is rich in saponins, a natural class of compounds used as a food additive for its amphiphilic properties, on the other hand during the hot water extraction several non-desirable compounds could be co-extracted.

What should be noticed is that using the same core material, supercritical clove oil under the same volume fraction of oily phase similar results were obtained in terms of emulsion droplet size approximately 0.2 μm (expressed as the Sauter, [D 3, 2]) using 100 g/ml of HICAP and only 0.02 g/ml, indicating the expressive surfactant properties of this alternative biosurfactant.

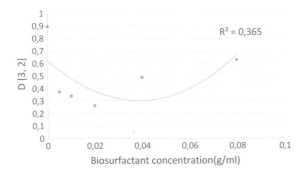

Fig. 5.6 Influence of Brazilian Ginseng roots extract concentration during ultrasound emulsification under pressure (ΔP of 0.75 atm) on mean droplet diameter (supercritical clove essential oil as oily phase, volume fraction of 2.5%)

Saponins are predominantly made of glycosides that possess one, two or three sugar chains attached to the aglycones, also called sapogenins, which are the nonpolar parts of the molecule [3]. The presence of hydrophobic and hydrophilic areas (the aglycone and sugar resides) in the saponin molecules account for the ability of this compound to reduce surface tension at phase boundaries [6]. These compounds have been used in foods as natural surfactants; they serve as preservatives to control the microbial spoilage of food. Because of consumer preferences for natural substance, saponins have more recently been used as a natural small molecule surfactant in beverage emulsions to replace synthetic surfactants such as polysorbates [7].

Based on the promising results regarding the use of aqueous extract from Brazilian ginseng roots as an alternative biosurfactant and the use of emulsification under nitrogen pressurized atmosphere we performed at the optimum biosurfactant concentration (0.02 g/ml), prepared dissolving the freeze-dried aqueous and hydrostatic pressure a study of the use of the an oily bixin-rich extract from annatto seeds (*Bixa orellana* L.) as core material obtained using as extracting solvent hot pressurized ethyl acetate. The organic solvent ethyl acetate was chosen because it demonstrated to be at least 2.3 times more selective than ethanol for bixin recovery using a lot of raw material, since using ethanol the bixin content in the solid extract under optimized conditions is 7.58 [8], meanwhile using ethyl acetate a value of 18.1 was obtained using nonoptimized extraction conditions [9].

On the other hand, since ethyl acetate has a toxicological classification that should be accountable for safety reasons the amount of it in the final food product we proposed the further processing of this emulsion by Supercritical Fluid Extraction of Emulsions (SFEE) process. Together with organic solvent elimination rate what can be observer is a good collateral effect regarding a fractionation/purification of bixin in the final suspension (product with residual amount of organic solvent). The analysis of the residual ethyl acetate concentration 9.4 ppm corroborating literature data, which found a concentration commonly lower than 50 ppm when supercritical antisolvent processes are used, while the conventional solvent evaporation results in a residual content of around 500 ppm [10]. Regarding the a possible fractionation/purification of bixin in the final suspension further studies will be done by our research group in addition to the economics aspects regarding supercritical antisolvent-based processes. On the other hand, regarding droplet size similar results were obtained for the emulsion (549 nm) and the produced suspension (569 nm), which were 24.74% lower that when no nitrogen atmosphere was used (730 nm).

SFEE combines the emulsion techniques and the SAS precipitation process. Emulsion techniques generally require large quantities of organic solvents, and their removal involves additional separation techniques and the use of high temperatures. In addition, SAS is not able to produce particles within the nanometric scale, and the resulting products have an increased tendency for particle agglomeration. To overcome these disadvantages, Chattopadhyay et al. [11] combined the two technologies and patented a new encapsulation method termed as Supercritical Fluid Extraction of Emulsions (SFEE). This process is also called Supercritical Emulsion Extraction (SEE). This method allows the removal of organic solvents during the process and enables the production of nanoscale particles that improve the solubility of the food or pharmaceutical solutes in aqueous solutions, which increases their bioavailability. Therefore, this work presents some important developments on the first step of the SFFE process, proposing instead of the conventional emulsification processes the use of pressurized nitrogen atmosphere to improve ultrasound emulsification during this step and consequently obtain SFEE products with lower droplet size. This proposed process we named as Ultrasound Emulsification Assisted by Nitrogen Hydrostatic Pressure (UEA-NHP) for SFEE.

5.4 Conclusions

In this work, the influence of hydrostatic pressure levels (up to 10 bars applying nitrogen), oily phase type, surfactant type, and surfactant concentration were evaluated. For all the oily phase tested, low over pressures (ΔP) of up to 0.5–0.75 bar improved the emulsification process, nevertheless, a further increase of pressure had a negative influence. Furthermore, when the pressure was elevated up to 2 bars, formation of emulsions stopped. Similar parabolic behavior was observed regarding surfactant concentration. Regarding, surfactant type, it can noticed that using the same core material, supercritical clove oil under the same volume fraction of oily phase similar results were obtained in terms of emulsion droplet size approximately 0.2 μm (expressed as the Sauter, [D 3, 2]) using 100 g/ml of HICAP and only 0.02 g/ml, indicating the expressive surfactant properties of nonpurified aqueous extract from *Pfaffia glomerata* roots.

In addition, the effect of the use of this alternative biosurfact and emulsification process was also evaluated to further processing of this emulsion by Supercritical Fluid Extraction of Emulsions (SFEE) process, using an oily bixin-rich extract from annatto seeds (*Bixa orellana* L.) as core material (extracting solution from hot ethyl acetate pressurized liquid extraction). Since, the final product of SFFE achieved a very low residual ethyl acetate concentration (9.4 ppm) and the regarding droplet size similar results were obtained for the emulsion (549 nm) and the produced suspension (569 nm), which were 24.74% lower that when no nitrogen atmosphere was used (730 nm). Therefore, we named as Ultrasound Emulsification Assisted by Nitrogen Hydrostatic Pressure (UEANHP) for SFEE (UEANHP-SFEE) this proposed combined process.

Acknowledgements Diego T. Santos thanks CNPq (processes 401109/2017-8; 150745/2017-6) for the post-doctoral fellowship. Ricardo A. C. Torres thanks Capes for their doctorate assistantship. M. Angela A. Meireles thanks CNPq for the productivity grant (302423/2015-0). The authors acknowledge the financial support from FAPESP (process 2015/13299-0).

References

1. S. Mitra, S.R. Dungan, Micellar properties of quillaja saponin. 2. Effect of solubilized cholesterol on solution properties. Colloids Surf. B **17**(2), 117–133 (2000)
2. E.K. Silva, M.T.M.S. Gomes, M.D. Hubinger, R.L. Cunha, M.A.A. Meireles, Ultrasoundassisted formation of annatto seed oil emulsions stabilized by biopolymers. Food Hydrocolloids **47**, 1–13 (2015)
3. W. Oleszek, A. Hamed, Saponin-based surfactants, in *Surfactants from Renewable Resources*, eds. by K. I. Johansson (John Wiley & Sons Ltd., United Kingdom, 2010), pp. 239–248
4. J.M. Prado, G.H.C. Prado, M.A.A. Meireles, Scale-up study of supercritical fluid extraction process for clove and sugarcane residue. J. Supercrit. Fluids **56**, 231–237 (2011)
5. O. Behrend, H. Schubert, Influence of hydrostatic pressure and gas content on continuous ultrasound emulsification. Ultrason. Sonochem. **8**(3), 271–276 (2001)
6. N. Mironenko, T. Brezhneva, T. Poyarkova, V. Selemenev, Determination of some surfaceactive characteristics of solutions of triterpene saponin derivatives of oleanolic acid. Pharm. Chem. J. **44**(3), 157–160 (2010)
7. C.Y. Cheok, H.A.K. Salman, R. Sulaiman, Extraction and quantification of saponins: a review. Food Res. Int. **59**, 16–40 (2014)
8. L.M. Rodrigues, S.C. Alcázar-Alay, A.J. Petenate, M.A.A. Meireles, Bixin extraction from defatted annatto seeds. C. R. Chim. **17**, 268–283 (2014)
9. D.T. Santos, M.T.M.S. Gomes, R. Vardanega, M.A. Rostagno, M.A.A. Meireles, Integration of pressurized fluid-based technologies for natural product processing, in *Natural Product Extraction: Principles and Applications*, eds. by M.A. Rostagno, J.M. Prado, RSC Green Chemistry No. 21 (2013)
10. D.T. Santos, D.F. Barbosa, K. Broccolo, M.T.M.S. Gomes, R. Vardanega, M.A.A. Meireles, Pressurized organic solvent extraction with on-line particle formation by supercritical anti solvent processes. Food Public Health **2**(6), 231–240 (2012)
11. P. Chattopadhyay, B.Y. Shekunov, J.S. Seitzinger, R. Huff, Particles from supercritical fluid extraction of emulsion. US Patent (2004)

Chapter 6
Economical Effects of Supercritical Antisolvent Precipitation Process Conditions

6.1 Introduction

Processes that employ supercritical carbon dioxide as a solvent have been successfully used by several researchers to micronize several food and pharmaceutical pure solutes due to its high solubility in supercritical CO_2. Kayrak et al. [1] and Pathak et al. [2] obtained micro- and nanoparticles via RESS (Rapid expansion of the supercritical solution) and RESOLV (Rapid expansion of a supercritical solution into a liquid solvent); the details of these processes have been reported by Sun et al. [3]. However, due to the low solubility of some high-added-value pure solutes used by the food and pharmaceutical industries, such as bixin and ibuprofen sodium in supercritical CO_2, for example, processes that use supercritical CO_2 as an antisolvent must be employed, such as SAS (Supercritical Antisolvent) and SFEE (Supercritical Fluid Extraction from Emulsions). Gomes et al. [4] have described these processes in detail.

Bakhbakhi et al. [5] and Martín et al. [6] successfully micronized ibuprofen sodium using the SAS process. Martín et al. [6] presented a study of the influence of different processes and operating parameters on the purity, particle size, morphology and polymorphism of ibuprofen sodium. Because the authors presented a limited parametric study, Bakhbakhi et al. [5] supplemented this study by increasing its range. Moreover, the "in vitro" drug performance was tested. The results obtained by the authors showed an improvement in the "in vitro" drug activity of the SAS-processed ibuprofen sodium.

The present work reports a systematic energetic-economic study of a supercritical CO_2-based micronization process using ibuprofen sodium as a model solute, addressing its industrial application through simulations performed to estimate the energy cost input required for the production of micronized food and pharmaceutical particles, such as ibuprofen sodium and bixin. The SAS process was used to re-crystallize the ibuprofen sodium salt from ethanolic solutions. Therefore, the

D. T. Santos et al., *Supercritical Antisolvent Precipitation Process*, SpringerBriefs in Applied Sciences and Technology, https://doi.org/10.1007/978-3-030-26998-2_6

effects of several operational parameters (pressure, temperature, CO_2 flow rate, solution flow rate, injector type and concentration of solute in the ethanol solution) on the energy cost per unit of manufactured product were investigated. In this work, two different injectors were used; a completely randomized experiment would eventually require a modification of the apparatus after each experimental run. To avoid this, the experimental runs were done accordingly with a split-plot experimental design [7].

6.2 Materials and Methods

6.2.1 Materials

Ibuprofen sodium salt (BCBC9914V, India) was purchased from Sigma-Aldrich and used as a model substance in the precipitation experiments. Ethanol (Dinâmica®, 52990, Diadema, Brazil), with a minimum purity of 99.5%, was used to prepare the ibuprofen sodium solutions. Carbon dioxide (99% purity, Gama Gases Especiais, Campinas, Brazil) was used as the antisolvent in the SAS process.

6.2.2 Experimental Procedure

A schematic diagram of the constructed experimental setup to perform the SAS precipitation experiments on a laboratory scale is shown in Fig. 6.1. The procedure was performed as follows: The CO_2 from the container is cooled to -10 °C using a thermostatic bath (Marconi, MA-184, Piracicaba, Brazil) to ensure the liquid CO_2 is being pumped by an the air-driven liquid pump (Maximator, M111 CO_2, Germany) in a 500 mL stainless steel (AISI 316) precipitation vessel with a 6.8 cm inner diameter. The precipitation vessel is fitted with an electric heating jacket and a AISI 316 stainless steel porous filter (screen size of 2 μm) fixed at the bottom of the vessel, which is used to collect the precipitated particles.

   Once the desired conditions of pressure, temperature, and CO_2 flow rate are achieved and remain stable, the ethanolic solution, which contains ibuprofen sodium, is introduced into the vessel by a high-performance liquid chromatography (HPLC) pump (Jasco, PU-2080, Japan), which allows a maximum working solution flow rate of 10 mL min^{-1}. A volume of 43 mL is injected into the precipitation vessel, and 10 mL of pure ethanol is then pumped to clean the tubes. Depending on the solution flow rate used, the time allowed for precipitation was 43 or 86 min.

   In this work, two different injectors were used to mix CO_2 and the solution at the inlet of the precipitation vessel. The injector is placed at the top of the precipitation vessel. A coaxial nozzle is used, which consists of a stainless steel tube with an inner diameter of 1/16 inch. (i.d. 177.8 mm) for the solution, placed inside a 1/8-in stainless

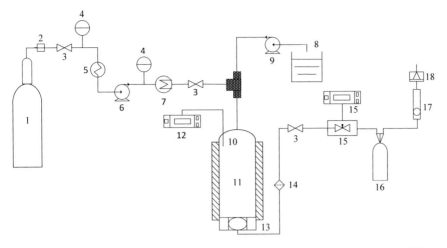

Fig. 6.1 Schematic diagram of the SAS apparatus. 1 CO_2 Cylinder; 2 CO_2 Filter; 3 Blocking Valves; 4 Manometers; 5 Thermostatic bath; 6 CO_2 Pump; 7 Heating bath; 8 Solution (solute/solvent) reservoir; 9 HPLC Pump; 10 Thermocouple; 11 Precipitation vessel; 12 Temperature controllers; 13 Filter; 14 Line filter; 15 Micrometric valve with a heating system; 16 Glass flask; 17 Glass float rotameter; 18 Flow totalizer

steel tube for the CO_2. A T-mixer is used, in which a 1/8-inch. stainless steel tube is used for both the solution and the CO_2. Figure 6.2 shows schematic diagrams of the two injectors.

When the solution and CO_2 are mixed, the ethanol is quickly solubilized by the supercritical CO_2, and this fluid mixture (CO_2 plus ethanol) exits the vessel and flows to a glass flask (100 mL) connected to a micrometric valve. This valve is maintained at 393 K to avoid the freezing and blockage of the outlet caused by the Joule–Thompson effect of the expanding CO_2. Ethanol is deposited in the glass flask, and the gaseous CO_2 is discharged to the atmosphere. The temperature and pressure were measured with instruments directly connected to the precipitation vessel with accuracies of ± 2 K and ± 0.2 MPa, respectively. The CO_2 flow rate is measured using a glass float

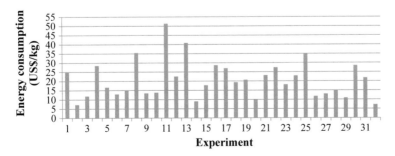

Fig. 6.2 Energy consumption per unit of manufactured product (US\$/kg) obtained in each experiment

rotameter (ABB, 16/286A/2, Warminster, USA) coupled with a flow totalizer (LAO, G0, 6, Osasco, Brazil).

After the injection of pure ethanol, the HPLC pump is stopped and only CO_2 is pumped, using a minimum of 300 g of CO_2 to ensure that all remaining traces of ethanol present in the precipitation vessel are removed before depressurization. Subsequently, the CO_2 pump is stopped, the precipitation vessel is slowly depressurized to atmospheric pressure and the particles are retained inside the precipitation vessel by a porous filter fixed at the bottom of the vessel and another placed at the vessel outlet (AISI 316 stainless steel porous line filter (Hoke, 6321G2Y, United States), porosity of 2 μm). The particles are carefully collected with a spatula and stored at ambient temperature in a glass desiccator and protected from light until subsequent analysis.

6.2.3   Determination of Energy Cost Per Unit of Manufactured Product

The energy consumption cost per unit of manufactured product was determined for each SAS experimental process condition using the SuperPro Designer 6.0® process simulator. This software allows the estimation of the mass and energy balance for all streams of the process. The results were normalized to determine the energy consumption (in terms of cost) per unit (1 kg) of manufactured product (particles obtained by SAS). The SAS process developed in the SuperPro Designer consisted of two pumps (one for CO_2 and one for ethanol), one precipitation vessel and two heat exchangers. The extraction procedure consisted of placing a known mass of ibuprofen sodium particles in contact with supercritical carbon dioxide plus ethanol.

6.2.4   Design of the Experiment and Statistical Analysis

The experiment was conducted using a split-plot experimental design. The injection type (T-mixer and coaxial nozzle) was applied to the whole plots with two replications and a 2^{5-2} fractional factorial while considering the temperature (313 and 323 K), pressure (10 and 12 MPa), concentration of ethanolic solutions (0.02 and 0.04 g mL^{-1}), CO_2 flow rate (500 and 800 g h^{-1}) and solution flow rate (0.5 e 1.0 mL min^{-1}) applied to the sub-plots, which totaled 32 experimental units. Treatments were deemed to be statistically significant when the p-value <0.1 (90% confidence limits). A statistical analysis was conducted with the MINITAB Statistical Software (Minitab Inc., State College, Pennsylvania). Table 6.1 shows the experimental conditions used for the SAS experiments randomized by split-plot design.

Table 6.1 Experimental conditions from split-plot design

Run	Injector	Temperature (K)	Pressure (MPa)	CO_2 flow rate (g h^{-1})	Solution flow rate (mL min^{-1})	Concentration of ethanolic solution (g mL^{-1})
1	T-mixer	313	12	800	1	0.02
2	T-mixer	313	10	500	0.5	0.04
3	T-mixer	313	12	500	1	0.04
4	T-mixer	323	10	500	1	0.02
5	T-mixer	323	12	800	0.5	0.04
6	T-mixer	313	10	800	0.5	0.02
7	T-mixer	323	10	800	1	0.04
8	T-mixer	323	12	500	0.5	0.02
9	Coaxial	323	12	500	0.5	0.04
10	Coaxial	313	12	800	1	0.04
11	Coaxial	323	10	800	1	0.02
12	Coaxial	313	10	500	0.5	0.02
13	Coaxial	323	12	800	0.5	0.02
14	Coaxial	323	10	500	1	0.04
15	Coaxial	313	10	800	0.5	0.04
16	Coaxial	313	12	500	1	0.02
17	Coaxial	313	10	800	1	0.02
18	Coaxial	323	10	800	0.5	0.04
19	Coaxial	313	10	500	1	0.04
20	Coaxial	323	12	800	1	0.04
21	Coaxial	313	12	800	0.5	0.02
22	Coaxial	323	10	500	0.5	0.02
23	Coaxial	313	12	500	0.5	0.04
24	Coaxial	323	12	500	1	0.02
25	T-mixer	323	10	800	0.5	0.02
26	T-mixer	313	12	800	0.5	0.04
27	T-mixer	323	12	500	1	0.04
28	T-mixer	323	12	800	1	0.02
29	T-mixer	323	10	500	0.5	0.04
30	T-mixer	313	12	500	0.5	0.02
31	T-mixer	313	10	500	1	0.02
32	T-mixer	313	10	800	1	0.04

6.3 Results and Discussion

The effect of the operating conditions on the energy cost per unit of manufactured product was also investigated while focusing on energy savings. Figure 6.2 shows the energy consumption (in terms of cost) per unit of manufactured product (US\$/kg) obtained in each experiment. The temperature and concentration of the ethanolic solution influenced energy consumption at statistically significant levels, as observed in Table 6.2. Furthermore, the statistical analysis indicated a second-order relationship between both parameters (p-value $= 0.097$) and between the pressure and solution flow rate (p-value $= 0.070$).

Figure 6.3 presents the concentration of the ethanolic solution as a function of temperature and the solution flow rate interactions as a function of pressure. The lowest estimated energy cost per unit of manufactured product was obtained for an ethanolic solution of 0.04 g mL^{-1} due to the higher manufactured solute mass. This result was independent of the temperature. Moreover, the lowest energy consumption was obtained at 12 MPa of pressure and a solution flow rate of 1 mL min^{-1}. The higher energy consumption obtained at a solution flow rate of 0.5 mL min^{-1}

Table 6.2 P-values obtained statistically for precipitation yield and residual solvent content

	Energy cost per unit of manufactured product
Parameter	p-value
Injector	0.147
Temperature	0.108
Pressure	0.894
CO_2 flow rate	0.578
Solution flow rate	0.589
Concentration of ethanolic solution	0.000

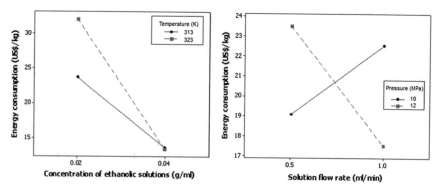

Fig. 6.3 Influence of Temperature × Concentration ethanolic solution and Pressure × Solution flow rate on energy consumption per unit of manufactured product (US\$/kg)

was due to a longer process. At 10 MPa, an opposite effect was observed due to the higher precipitation yield obtained at this solution flow rate, which reduced the energy consumption cost per unit of a manufactured product.

The cost of obtaining micronized solute particles is compensated by an improved solute bioavailability [1, 8]. The lowest energy consumption cost per unit of manufactured product (US$/kg) was obtained in experiments 2 and 32. Small micro- and nanometer-sized particles have attracted growing interest in the pharmaceutical and food industries because they endow materials with new properties that can be adopted by these industries [9]. Notably, the energetic cost (cost of utilities) is one of five factors used to estimate the cost of manufacturing (COM) according to a methodology proposed to Turton et al. [10]. The COM is estimated as the sum of the cost of investment, the cost of operational labor, the cost of the raw material, the cost of waste treatment and the cost of utilities (energetic cost). More information about each cost factor can be found in a paper by in Rosa and Meireles [11].

6.4 Conclusions

The effect of temperature versus concentration of ethanolic solution and pressure versus solution flow rate interactions SAS micronization on the energy consumption cost per unit of manufactured product was demonstrated, being the lowest estimated energy cost per unit of manufactured product was obtained using an ethanolic solution of 0.04 g mL^{-1} at 12 MPa and solution flow rate of 1 mL min^{-1}. This result was independent of the temperature. Thus, the present work increases knowledge about the energetic-economic aspects of the supercritical antisolvent micronization process, contributing to its further incorporation by the food and pharmaceutical industries.

Acknowledgements Diego T. Santos thanks CNPq (processes 401109/2017-8; 150745/2017-6) for the post-doctoral fellowship. Ricardo A. C. Torres thanks Capes for their doctorate assistantship. Juliana Q. Albarelli thanks FAPESP (processes 2013/18114-2; 2015/06954-1) for the post-doctoral fellowships. M. Angela A. Meireles thanks CNPq for the productivity grant (302423/2015-0). The authors acknowledge the financial support from FAPESP (process 2015/13299-0).

References

1. D. Kayrak, U. Akman, Ö. Hortaçsu, Micronization of ibuprofen by RESS. J. Supercrit. Fluids **26**(1), 17–31 (2003)
2. P. Pathak, M.J. Meziani, T. Desai, Y.-P. Sun, Formation and stabilization of ibuprofen nanoparticles in supercritical fluid processing. J. Supercrit. Fluids **37**(3), 279–286 (2006)
3. Y.P. Sun, M.J. Meziani, P. Pathak, L. Qu, Polymeric nanoparticles from rapid expansion of supercritical fluid solution. Chem.-A Eur. J. **11**(5), 1366–1373 (2005)
4. M.T.M.S. Gomes, D.T. Santos, M.A.A. Meireles, Trends in particle formation of bioactive compounds using supercritical fluids and nanoemulsions. Food Public Health **2**(5), 142–152 (2012)

5. Y. Bakhbakhi, S. Alfadul, A. Ajbar, Precipitation of Ibuprofen Sodium using compressed carbon dioxide as antisolvent. Eur. J. Pharm. Sci. **48**, 30–39 (2013)
6. Martín An, K. Scholle, F. Mattea, D. Meterc, Cocero MaJ, Production of polymorphs of ibuprofen sodium by supercritical antisolvent (SAS) precipitation. Crystal Growth Des. **9**(5), 2504–2511 (2009)
7. P. Huang, D. Chen, J.O. Voelkel, Minimum-aberration two-level split-plot designs. Technometrics **40**(4), 314–326 (1998)
8. M. Newa, K.H. Bhandari, J.O. Kim, J.S. Im, J.A. Kim, B.K. Yoo, J.S. Woo, H.G. Choi, C.S. Yong, Enhancement of solubility, dissolution and bioavailability of ibuprofen in solid dispersion systems. Chem. Pharm. Bull. **56**(4), 569–574 (2008)
9. P. Sanguansri, M.A. Augustin, Nanoscale materials development—a food industry perspective. Trends Food Sci. Technol. **17**(10), 547–556 (2006)
10. R. Turton, R.C. Bailie, W.B. Whiting, J.A. Shaeiwitz, *Analysis, Synthesis and Design of Chemical Processes*, 2th edn. (Pearson Education, 2008)
11. P.T. Rosa, M.A.A. Meireles, Rapid estimation of the manufacturing cost of extracts obtained by supercritical fluid extraction. J. Food Eng. **67**(1), 235–240 (2005)

Printed in the United States
By Bookmasters